STOP MANAGING CHANGE LIKE IT'S 2015

OCM FOR OUR AI-FIRST REALITY

AREND SCHURING

ANDY FORBES

Roadmap Press
Hillsboro, Virginia, United States

ISBN: 978-1-969489-13-6 (paperback)
ISBN: 978-1-969489-10-5 (ebook)

Cover design by: Roadmap Press

To my husband, Thomas McClure, for being my partner in every transformation. And to my mom, Jane, and my sister, Savannah, for teaching me resilience and proving that ideas are worth sharing. Congratulations on your own books; your journeys paved the way for mine.
Arend Schuring

To the colleagues and clients I've worked alongside over the years, thank you for teaching me, again and again, how demanding real change can be, and how transformative it becomes when people commit to doing it well. Your challenges, your candor, and your partnership have shaped every insight in these pages.
Andy Forbes

CONTENTS

PREFACE: NAVIGATING THE AI REVOLUTION

Welcome to Stop Managing Change Like It's 2015. This book arrives at an inflection point in business history: when artificial intelligence transforms from competitive advantage to existential necessity. Organizations now face critical questions that transcend technology: How must we reinvent ourselves to thrive in an era where human-AI collaboration becomes the primary source of value creation?

The coming decade will witness unprecedented organizational evolution. Traditional structures built around hierarchical control, fixed roles, and exclusively human decision-making will give way to fluid collaborative networks where humans and AI systems form dynamic partnerships based on comparative advantage. Leaders who recognize the depth of this transformation will help shape their industries while those who treat AI as just another technology initiative risk joining the corporate fossil record.

This book provides a comprehensive framework for navigating this transformation through practical, actionable guidance grounded in real-world implementation.

HOW TO USE THIS BOOK

This book is designed to support multiple audiences navigating AI-driven transformation from different vantage points. While some readers may choose to read it cover to cover, others will benefit most from focusing on the chapters aligned to their role and responsibilities.

For CEOs and Executive Teams: Begin with Chapters 1–3 to establish the strategic context for AI-driven transformation and the leadership shifts it requires. Then focus on Chapters 10–12, which address human-first governance, external ecosystem dynamics, and how to sustain momentum as AI reshapes organizations and industries.

For Functional Leaders: Chapters 4–8 explore how AI changes work, roles, operating models, and functional execution, supported by practical frameworks and real-world scenarios. Chapter 9 is recommended for leaders responsible for translating strategy into coordinated execution across teams.

For HR and Change Management Professionals: The core change and adoption content spans Chapters 4–5, 7–9, and Chapter 10. These chapters address capability building, behavioral adoption, practical change frameworks, implementation realities, and human-centered governance in AI-enabled organizations.

For Technology Leaders: While this book focuses on organizational rather than technical implementation, Chapters 2, 6, 9, and 11 provide valuable perspective on how AI reshapes work, industry patterns, execution models, and the broader AI landscape. The appendices offer additional guidance for aligning technology decisions with organizational transformation goals.

THROUGHOUT THIS BOOK

Throughout the book, you will find several practical elements designed to help translate ideas into action:

Real-World Scenarios and Illustrative Examples: Composite scenarios drawn from organizational experiences show how AI-driven change plays out across industries, functions, and roles, highlighting both effective practices and common missteps.

Applied Guidance and Implementation Insight: Rather than abstract theory, the chapters emphasize how leaders and teams can apply concepts in practice, from redesigning work and roles to navigating adoption challenges and governance decisions.

Patterns, Signals, and Trade-Offs: Across chapters, recurring patterns and decision points are surfaced to help readers recognize what matters most as AI reshapes work, culture, and organizational design.

Chapter Summaries and Key Takeaways: Each chapter concludes with concise takeaways that reinforce the most important insights and clarify how they connect to broader transformation efforts.

Most importantly, this book treats AI transformation as a human endeavor. The practical elements are intended not as prescriptive templates, but as tools for sensemaking, helping leaders and practitioners interpret complexity, make better decisions, and adapt as conditions continue to evolve.

The journey ahead will challenge every assumption about how organizations function. It will demand courage to question established structures, leadership humility to acknowledge uncertainty, and unwavering focus on enhancing human capability alongside technological advancement.

One note: This book draws on publicly available information, industry reporting, and the author's professional experience. References to companies, platforms, or individuals are illustrative and based on public statements and observable market behavior, not on non-public information or insider knowledge.

We invite you to approach this material not as a prescriptive template but as a strategic dialogue that will evolve with your own implementation experience. The future belongs to those who recognize that humanizing the swarm (creating organizational forms where humans and AI together achieve what neither could alone) represents the ultimate competitive advantage.

1

EMBRACING THE AI SWARM

When we speak of the 'swarm,' we do not mean a mindless collective. In a machine swarm, every node is identical. In a human-AI swarm, every human node adds unique value (such as context, ethics, and empathy) that the network cannot generate on its own. The goal is not to assimilate humans into the machine, but to use the machine's connectivity to amplify human intent.

The revolution unleashed by artificial intelligence demands nothing less than complete organizational metamorphosis. The transformation is fundamentally reimagining how work gets done, decisions get made, and value gets created.

Most AI change efforts are still framed as technical upgrades, workflow redesigns, or employee reskilling initiatives. While important, these are first-order actions. The more consequential challenge is enabling organizations to become adaptive knowledge systems. Most organizations approach AI adoption with the same energy they apply to cleaning out the office fridge: they know it's necessary, they know it's going to be unpleasant, and they really hope someone else does it. We label these efforts 'workflow redesigns,' which is often a euphemism for 'digitizing the chaos.' If you automate a broken process, you don't get efficiency; you just get to the disaster significantly faster.

In a landscape where AI continuously generates, summarizes, and recombines insights, advantage no longer stems from holding proprietary knowledge but from how quickly an organization can absorb external knowledge, synthesize it with internal context, and redeploy it into action. Knowledge agility becomes the new currency.

To lead in this era, organizations must treat knowledge flow as an enterprise capability. That includes:

- Building knowledge capture mechanisms that surface insight from human-AI collaboration.
- Enabling dynamic learning networks across departments and ecosystems, not just static centers of excellence.
- Creating real-time sensemaking practices that help teams adapt to volatility, interpret weak signals, and share interpretations.

This reframing requires updating not only our processes but our diagnostic models. Traditional assessments look at leadership alignment, but in the AI era, we must assess Knowledge Ecosystem Readiness. Ask yourself: How fast do ideas move across your silos? In many organizations, information travels between departments with the speed and aerodynamic efficiency of a sloth moving through wet concrete. If your marketing team's brilliant insight takes three quarters, five committees, and a blood oath to reach product development, your 'ecosystem' isn't a network; it's a series of disconnected fortresses protecting their budgets. Organizational Change Management must now include practices that turn knowledge velocity into strategic leverage. Change leaders are now designing how insight travels and transforms across the enterprise.

Consider how one global manufacturer abandoned traditional hierarchies for AI-augmented teams. Their maintenance division morphed from supervisor-heavy chains of command into self-organizing units where AI monitors equipment, predicts failures, and coordinates human expertise precisely where needed. Productivity increased while employee satisfaction scores. The secret? The organizational architecture that enabled humans and AI to collaborate seamlessly.

The transformation unfolds across multiple dimensions:

Structural Evolution: Rigid hierarchies collapse into network architectures where teams form, execute, and dissolve based on emerging needs. At one financial firm, traders now work alongside AI systems that continuously analyze market patterns, suggesting opportunities that humans evaluate and execute within minutes rather than days. This hybrid structure shortened decision cycles by 60 percent while improving risk assessment.

Process Reinvention: Static workflows dissolve into dynamic task allocation based on comparative advantage. A shipping company transformed package delivery by decomposing routes into optimizable components, then using AI to handle sequencing while reserving customer relationship management for human drivers. The result is a fundamentally different operational model that competitors struggle to replicate.

Governance Transformation: Command-and-control oversight gives way to principle-based guardrails. A video on demand company abandoned detailed content production guidelines in favor of AI-enhanced decision frameworks that balance creative freedom with data-driven insights. Their production teams achieve both artistic excellence and commercial success because the system constantly learns and adapts without stifling human creativity.

Human Renaissance: Perhaps most profound (and least understood) is how AI liberates human potential. As machines take on routine analysis, humans focus on innovation, empathy, and complex judgment. A software company's customer service transformation created AI systems that handle 78 percent of routine inquiries while elevating human agents to relationship managers who address complex problems. Agent satisfaction increased dramatically as their work became more meaningful.

It has been publicly reported that John Chambers, former Cisco CEO, observed that some companies won't survive the next decade not because they lack technology, but because they fail to transform their organizational DNA to thrive in an AI-powered world. The scope, speed, and nature of AI transformation demand frameworks built specifically for this revolution. Success requires leaders who embrace uncertainty, model vulnerability, and demonstrate adaptability at every turn. AI provides the catalyst, but the alchemy happens when organizations reimagine themselves around human-AI synergy. Those who grasp this will dominate. Those who don't will join the corporate fossil record, potentially creating excellent oil reserves for the AI-powered civilizations of the 23rd century. It's a legacy, certainly, just not the one your shareholders are hoping for.

THE COMING TRANSFORMATION: WHY AI DOMINANCE IS INEVITABLE

The confluence of technological, economic, and societal forces is turning AI from competitive edge to survival necessity. Organizations now face a strategic inflection point that redefines winners and losers.

The convergence of technological breakthroughs, economic imperatives, and societal shifts has created an environment where AI adoption has shifted from competitive advantage to survival necessity. Unlike previous technological waves that allowed for gradual adaptation, AI's transformation curve has turned nearly vertical - reshaping entire industries before many incumbents have even formulated a response.

The evidence surrounds us in three converging forces:

Technological Acceleration: Neural networks now routinely outperform humans across expanding cognitive domains, from medical diagnosis to legal analysis to creative generation. Each breakthrough unlocks cascading possibilities as capabilities compound rather than simply add. Consider language models: in just 24 months, they evolved from crude text completion to sophisticated reasoning engines capable of complex problem-solving. It has been reported that Microsoft's Chief Scientist compared this acceleration curve to watching evolution compressed into months rather than millennia.

Economic Inevitability: Organizations leveraging AI are achieving transformative performance improvements in specific operations, not just marginal gains. By automating complex, document-heavy workflows, AI and machine learning allow businesses to fundamentally reinvent processes. The result is a new economic structure for those functions, enabling a reduction in processing times and significant improvements in quality, customer satisfaction, and scalability.

The competitive divide created by AI is accelerating. Early adopters are building compounding advantages that will fundamentally reshape market dynamics. While the precise metrics vary by industry, analyses consistently show that the most advanced AI users are experiencing disproportionate productivity gains and stronger financial performance. As AI innovation races ahead and investment intensifies, the gap between early adopters and laggards will continue to widen. For businesses that fail to integrate AI effectively, catching up may prove nearly insurmountable within the next few years.

Societal Transformation: Consumer expectations have fundamentally shifted. Interactions that aren't instant, personalized, and predictive now register as substandard service. B2B buyers expect AI-enhanced solutions. The labor market increasingly rewards AI-augmented capabilities. The marketplace itself is voting for AI through consumption choices, talent flows, and investment patterns.

Traditional organizational structures built on static roles, fixed hierarchies, and human decision-making become competitive liabilities in AI-powered markets. The organizations thriving in this environment have decomposed work into its fundamental components, enabling dynamic allocation between human and machine intelligence based on comparative advantage.

Consider public reporting on how Ping An Insurance transformed claims processing. Rather than automating existing workflows, they decomposed the entire process into many distinct tasks. AI systems dynamically route each component to the optimal resource: simple damage assessment to computer vision models, complex liability questions to human adjusters, customer communication handled through hybrid interfaces. The system continuously improves as it learns which allocation patterns produce the best outcomes.

This shift from role-based to task-based organization represents organizational evolution in real-time. Every position decomposes into discrete, measurable activities. Each undergoes continuous ROI evaluation to determine optimal execution: human, AI, or collaborative. Human work evolves toward uniquely human domains (i.e., creativity, empathy, ethical judgment, and strategic thinking) while machines handle analytical heavy lifting. This is the essence of humanizing the swarm. We do not decompose jobs simply to optimize costs; we do it to isolate the 'robotic' drudgery that stifles human creativity. By handing the repetitive tasks to the swarm, we clear the space for what we call Synthetic Self-Actualization - the use of AI to accelerate an employee's journey toward their highest professional contribution. By liberating employees from routine tasks, AI enables them to pursue work aligned with their uniquely human capabilities. Companies that intentionally design this transition create unprecedented employee engagement while simultaneously driving performance improvements.

Consider pharmaceutical company's approach to drug development. Their AI systems handle computational tasks that previously consumed 70 percent of researcher time. Scientists now focus exclusively on creative hypothesis generation and complex judgment. The result is dramatically higher retention of top talent who report greater fulfillment in their now-augmented roles.

The window for proactive transformation is closing rapidly.

FROM TASK DECOMPOSITION TO KNOWLEDGE RECOMPOSITION

Static job descriptions increasingly constrain both operational performance and human potential. Organizations that decompose roles into optimizable tasks unlock unprecedented opportunities for both efficiency and fulfillment.

Task decomposition is a systematic methodology that transforms monolithic roles into dynamic task portfolios, enabling continuous optimization based on analysis of comparative advantage between humans and machines. The brutal truth most executives avoid confronting: 40-70 percent of current organizational roles contain tasks that algorithms could perform better, faster, and more consistently than humans.

In traditional process transformation, task decomposition is used to break work into smaller, standardized components, each eligible for optimization, automation, or outsourcing. This logic still applies in the AI age, but its value has shifted. More than productivity, the goal is *knowledge visibility*. By decomposing workflows, organizations can unbundle where knowledge lives: who interprets it, who transforms it, where it's stuck, and how it evolves. Every decomposed task becomes a diagnostic lens for spotting tacit judgment, contextual awareness, and cognitive leaps, things AI can't replicate without human scaffolding. This opens the door to knowledge recomposition: reconfiguring how insight is produced, transferred, and reused. That might mean:

- Embedding AI in places where decision support is needed.
- Redesigning workflows to make human interpretation more visible and teachable.
- Shifting roles from task execution to knowledge synthesis and guidance.

Task decomposition is the starting point for designing adaptive knowledge systems. What matters is not how many tasks can be handed off to AI, but how well humans and AI can co-evolve knowledge in ways that compound over time.

Task decomposition follows a methodology consisting of three phases:

Granular Analysis: Every role undergoes systematic dissection into discrete, measurable tasks. Each component requires precise definition of inputs, outputs, decision points, dependencies, and success metrics. When health insurance company decomposed their claims processing roles, they identified 2,300 distinct tasks across what had previously been viewed as a handful of job descriptions. Task decomposition is a brutal exercise. It forces you to look at a highly paid executive's day and realize that 60% of their 'strategic oversight' is actually just 'forwarding emails with the subject line: Thoughts???' By unbundling these tasks, we can finally let AI handle the 'Thoughts???' emails, freeing the human to actually have some thoughts.

Strategic Evaluation: Each task undergoes assessment across multiple dimensions:

- Automation potential: technical feasibility and implementation complexity
- Value contribution: impact on customer experience, revenue, and strategic objectives
- Human comparative advantage: creativity, judgment, and emotional intelligence requirements

- Risk parameters: error tolerance and regulatory considerations

Allocation Optimization: The insights from evaluation drive dynamic task allocation:

- Tasks with >30 percent ROI improvement through AI: automate completely
- Tasks requiring human judgment/creativity but benefiting from AI support: augment
- Tasks demanding emotional intelligence: preserve for humans
- Tasks requiring hybrid capabilities: design collaborative workflows

The resulting transformation reshapes work profoundly. Consider financial analysis roles: Traditional analysts wasted 60 percent of their time on data gathering, report generation, and routine calculations. Post-decomposition, AI handles these components with greater speed and accuracy, while humans focus on pattern interpretation, strategic recommendations, and stakeholder communication.

This pattern repeats across every function:

Sales: AI manages lead scoring, initial outreach, and CRM updates while humans focus on relationship building, complex negotiations, and strategic account planning. One pharmaceutical company increased sales force effectiveness 35 percent through this rebalancing.

HR: AI handles resume screening, benefits administration, and routine inquiries while humans concentrate on culture development, talent strategy, and complex employee situations. A software company reduced recruitment cycles by 60 percent while improving candidate satisfaction through this approach.

Marketing: AI optimizes campaigns, segments audiences, and analyzes performance while humans drive creative strategy, brand storytelling, and market positioning. A video on demand company attributes their content strategy success to this division of cognitive labor between algorithms and creative executives.

The implications cascade throughout organizations:

- Departmental boundaries blur as tasks recombine around optimal execution patterns
- Management shifts from supervision to orchestration as AI handles coordination functions
- Decision velocity accelerates through AI-powered analytics and simulation capabilities
- Organizational structure flattens as information asymmetries collapse

By automating routine tasks across departments AI is forcing a re-examination of core organizational tenets. This reallocation of cognitive labor creates a new blueprint for the enterprise, with profound downstream effects on how work is structured and executed:

Ruthless Objectivity: Political considerations and tradition cannot influence task evaluation.

Employee Partnership: Successful decomposition requires active workforce participation. Workers understand nuances invisible to management and their involvement builds essential buy-in.

Continuous Refinement: Task portfolios evolve as AI capabilities advance. Organizations need systematic review processes to reassess allocation as technology evolves.

Human-Centric Design: Optimization must consider employee fulfillment, not just efficiency. The goal is to elevate work to its highest form. Organizations that design for meaningful human contribution alongside machine efficiency create sustainable advantage through both performance and talent attraction.

Organizations mastering task decomposition gain compound advantages. They achieve operational excellence while simultaneously enhancing employee engagement. They build adaptable structures that evolve with AI advancement. Most importantly, they position themselves to continuously outpace competitors trapped in role-based thinking.

AI PROCESS ENGINEERING – DESIGNING THE NEW WORK STRUCTURE

Task decomposition provides the blueprint, but process engineering brings it to life. Organizations that orchestrate the intricate dance between human and machine intelligence create capabilities their competitors cannot easily replicate.

Process engineering creates the architecture that orchestrates the complex dance between human and machine intelligence. While decomposition identifies what to optimize, process engineering determines how these components reconnect into coherent workflows. Organizations mastering this discipline fundamentally rewire their operational DNA.

A video on demand company revolutionized content creation by engineering entirely new processes rather than simply automating existing ones. They created integrated workflows where AI continuously analyzes viewer preferences, predicts emerging interests, and provides creative teams with insights that shape production decisions. This is a symbiotic partnership between human creativity and machine intelligence that has transformed the economics of entertainment production.

Process engineering begins with architectural vision that spans four domains:

Workflow Mapping: Unlike traditional process documentation focused on sequential handoffs, AI-enabled workflows require multi-dimensional mapping. Every process undergoes analysis to identify value streams, constraint points, and decision nodes where AI and humans intersect.

Task Optimization: The insights from decomposition drive workflow redesign that integrates AI at high-impact points. A health organization engineered new diagnostic processes where AI systems evaluate medical images and patient data, flagging potential issues for physician review. This hybrid approach improved diagnostic accuracy by 30 percent while reducing physician burnout by eliminating routine reviews. The handoff protocols ensure physicians maintain ultimate responsibility while AI handles data-intensive analysis.

System Integration: Forward-thinking companies are cutting costs and speeding up projects by creating a library of AI tools that any team can use, rather than reinventing the wheel for every new application.

Governance Framework: Because AI is being built into our systems, we must have clear rules to manage it. This requires setting up processes that assign responsibility, measure performance, monitor for issues, and have protocols for problems, which builds essential trust.

The sophistication lies in the interfaces. Unlike crude automation that simply replaces humans, AI process engineering creates intelligent collaboration. Consider customer service transformation at American Express: AI provides agents with real-time sentiment analysis, suggested responses, and complete interaction history. When complex issues arise, the transition from AI to human happens seamlessly, with perfectly prepared handoffs that enable agents to deliver exceptional service.

The strategic deployment of AI hinges on a methodical, multi-phase approach that builds from a solid foundation to full-scale innovation:

Phase 1: Foundation

- Map current-state processes with ruthless honesty about inefficiencies
- Complete task decomposition analysis to identify optimization opportunities
- Design initial human-AI interfaces with careful attention to handoff points
- Build pilot programs in controlled environments to validate architectural assumptions

Phase 2: Integration

- Deploy AI systems incrementally with careful monitoring of both technical and human factors
- Train human teams with emphasis on collaboration skills, not just technical operation
- Refine handoff protocols based on real-world interaction patterns
- Measure performance impacts across both efficiency and experience dimensions

Phase 3: Optimization

- Scale successful implementations with appropriate customization for local conditions
- Enhance AI capabilities through continuous learning from operational data
- Evolve human roles toward higher-value activities as AI demonstrates reliability
- Capture compound benefits as human-AI teams develop new capabilities

Phase 4: Innovation

- Experiment with emerging AI capabilities in controlled environments
- Create entirely new service offerings enabled by human-AI collaboration
- Develop competitive advantages based on proprietary process patterns
- Foster continuous evolution of both technology and organizational capabilities

Beyond the technical rollout, sustainable AI success is built upon a robust governance model that is designed to evolve alongside implementation, ensuring control, managing risk, and driving continuous improvement:

Performance Management: Dashboards track both technical metrics (accuracy, processing time, exception rates) and human factors (satisfaction, adoption, contribution quality). These balanced scorecards prevent the common failure of optimizing algorithms at the expense of human experience.

Risk Mitigation: Comprehensive monitoring identifies potential issues before they impact operations. Proactive controls include bias detection, input validation, and exception handling protocols. Progressive implementation with appropriate human oversight manages transition risks effectively.

Continuous Improvement: Regular process audits compare actual performance against expected outcomes. Feedback loops capture insights from both AI analytics and human experience. Knowledge management systems ensure learning spreads throughout the organization.

Organizations that excel at AI process engineering create self-reinforcing advantages. Efficiency gains fund further AI investments. Enhanced employee satisfaction attracts top talent. Operational flexibility enables rapid market response. The gap between leaders and laggards widens exponentially with each implementation cycle.

Yet the human dimension (i.e., managing resistance, building readiness, fostering engagement) determines ultimate success. This critical challenge demands equal sophistication and strategic attention.

FROM RESISTANCE TO READINESS – CONVERTING FEAR INTO COMPETITIVE ADVANTAGE

Employee resistance to AI contains valuable intelligence that most organizations inadvertently suppress. Leaders who transform this natural skepticism into implementation insight create momentum their competitors can't match.

The case for AI transformation is clear: the technology delivers, and the economics make sense. Yet most initiatives fall short of expectations. The culprit isn't technical complexity, but a failure to lead the human side of change. Organizations that embrace the psychology of transformation gain a lasting edge by accelerating execution and unlocking real value.

Employees resist AI implementation because they see threats to job security, professional identity, and personal agency. They question whether leadership truly understands the operational implications. Leaders who dismiss these concerns as irrational opposition create underground resistance that sabotages transformation. Leaders who treat resistance as valuable feedback create passionate change advocates and uncover critical implementation blind spots.

The psychology of AI resistance follows predictable patterns that require specific leadership responses:

Stage 1: Denial ("AI won't affect my role")

- Provide concrete evidence of AI's capabilities and limitations
- Share industry transformation data showing competitive pressures
- Demonstrate AI applications through hands-on exposure

- Create safe spaces for questions and concerns without judgment

Stage 2: Fear ("AI will replace me")

- Articulate vision of human-AI collaboration, not replacement
- Provide concrete reskilling pathways with resource commitments
- Guarantee job security during transition period where feasible
- Address status and identity concerns, not just employment fears

Stage 3: Bargaining ("Maybe AI can help with some tasks")

- Launch pilot programs with volunteers from skeptical groups
- Celebrate early wins publicly with emphasis on participant experiences
- Share productivity and satisfaction improvements with concrete metrics
- Allow early adopters to become evangelists rather than forcing top-down messaging

Stage 4: Exploration ("How can I work effectively with AI?")

- Provide comprehensive training programs focusing on collaborative skills
- Create AI literacy curricula tailored to different roles and technical backgrounds
- Establish mentorship networks connecting early adopters with newcomers
- Reward experimentation and learning

Stage 5: Commitment ("AI makes me more valuable")

- Recognize AI collaboration in performance reviews and career advancement
- Create new career paths incorporating emerging AI-related skills
- Share success stories showcasing enhanced human capabilities
- Build culture celebrating human-AI partnership as the new standard of excellence

To shift mindsets and behaviors at scale, organizations must deploy targeted strategies that move employees from initial resistance to active advocacy:

Communication Architecture: Change succeeds when messaging is clear, consistent, and tailored. A one-size-fits-all announcement breeds skepticism. Instead, use multiple channels and customize messages for different stakeholder groups to build trust and engagement.

Capability Building: Start with organization-wide AI literacy to create a shared understanding. Then, offer role-specific training that builds confidence through practical application. Equip leaders with tools to manage AI-integrated teams and navigate the complexities of human-machine collaboration.

Support Systems: Transformation is both technical and emotional. Provide access to coaches who can guide employees through new workflows and mindset shifts. Foster peer networks to share experiences and offer mutual support. Help employees plan career transitions with clear pathways for evolving roles.

Cultural Reinforcement: Lasting change requires cultural integration. Update values to reflect human-AI collaboration. Recognize and reward successful adaptation. Share stories of transformation to inspire others. Ultimately, make AI-augmented work part of the organization's identity.

While strategic interventions lay the foundation for change, it is leadership behavior that ultimately determines whether transformation efforts succeed or stall:

Radical Transparency: Effective change requires complete honesty about implications.

Genuine Empathy: Change leaders must acknowledge emotional impacts before attempting rational arguments.

Consistent Action: Leaders must align decisions with stated values throughout implementation.

Visible Leadership: Executives must personally demonstrate engagement with AI transformation.

Patient Persistence: Cultural transformation requires 18-36 months of sustained effort. Leaders must maintain momentum through inevitable challenges without declaring premature victory.

Organizations that prioritize human transformation unlock enduring competitive advantage.

Employees shift from passive participants to active innovators, often developing AI applications beyond leadership's initial vision. Resistance gives way to resilience. Fear evolves into enthusiasm. The organization becomes fundamentally more capable and adaptive.

This human-centered foundation sets the stage for the next essential shift: leadership transformation. To lead effectively in an AI-enabled environment, executives must reimagine their roles, moving from traditional oversight to orchestrating seamless collaboration between humans and machines.

LEADING WITH THE SWARM – THE EXECUTIVE'S ORCHESTRATION PLAYBOOK

The leadership capabilities that built today's successful organizations increasingly constrain tomorrow's possibilities. Executives who evolve from commanders to orchestrators unlock collective intelligence at scales previously unimaginable.

The move from hierarchical control to swarm orchestration represents a historic shift in leadership. Executives who once excelled in command-and-control environments now face a new challenge: driving performance in dynamic human-AI ecosystems where traditional authority structures no longer apply.

The new leadership imperatives demand fundamental shifts in executive mindset and practice:

From Efficiency to Ecology: The swarm naturally optimizes for speed. The leader's new role is to act as the ecosystem's conscience. You must ask not just "Is the swarm working fast?" but "Is the swarm serving the people?" Leadership provides the moral guardrails that algorithms cannot invent.

From Control to Choreography: Successful AI-era leaders design systems where optimal decisions emerge organically rather than through hierarchical approval. They establish boundary conditions rather than micromanaging steps. This orchestration approach creates environments where human and machine distributed intelligence converges on superior solutions.

From Answers to Architecture: Leaders must design learning systems that continuously generate insights. The executive's value comes from asking the right questions and creating conditions for discovery rather than providing answers.

From Management to Meaning: As AI handles increasing portions of operational management, leadership's most important function becomes cultivating purpose that goes beyond efficiency. Executives must connect individual growth to organizational success and transform work into a vehicle for self-actualization.

To lead effectively in AI-powered, decentralized environments, executives must evolve beyond traditional leadership models. These five capabilities to inspire alignment, enable distributed decision-making, and sustain momentum in complex, fast-moving systems:

Strategic Narrative Mastery: Leaders must craft compelling stories that give AI transformation meaning beyond efficiency. These narratives connect technological change to fundamental human values and paint vivid pictures of enhanced futures.

Radical Transparency: Trust becomes the currency enabling bold AI moves. Leaders build this trust through unprecedented openness: sharing decision-making logic, acknowledging uncertainties, providing access to performance data, and creating psychological safety for difficult conversations.

Distributed Intelligence Architecture: Effective leaders design systems that leverage collective wisdom rather than funneling decisions through hierarchical channels. They enable decision-making at optimal levels and create feedback loops for continuous learning.

Cultural Amplification: AI-era leaders recognize that culture becomes their primary competitive advantage. They use technology to reinforce desired behaviors, measure cultural health, and create environments where values become operational rather than aspirational.

Adaptive Capacity Building: The most critical leadership function becomes developing organizational adaptability. Leaders model personal learning, invest in capability development, and create structures that enable rapid evolution.

The daily rhythm of swarm leadership looks nothing like traditional executive routines.

In AI-enabled organizations, leaders shift from oversight to orchestration, guiding dynamic systems through insight, engagement, and adaptability.

- Instead of scanning static performance reports, swarm leaders begin their day with AI-generated insights on organizational health. They apply human judgment to set strategic priorities and communicate decisions with full context and transparency.
- They engage directly with human-AI teams. They observe collaboration patterns, identify systemic improvements, and remove barriers to swarm effectiveness.
- Moving beyond exception management, leaders partner with AI to explore long-term scenarios. They blend data-driven analysis with intuition and craft narratives that connect today's actions to the organization's broader purpose.
- The day ends not with a focus on results alone, but with reflection. Leaders review metrics on their own effectiveness, set personal learning goals, and design interventions to strengthen the organization's adaptive capacity.

True transformation begins with the leader.

Before an organization can evolve, its executives must undergo their own shift by building the mindset, skills, and behaviors needed to lead in an AI-powered, adaptive environment.

Phase 1: Personal Transformation

Leaders must first embrace change themselves:

- Build executive-level AI literacy through immersive learning experiences.
- Demonstrate vulnerability by openly acknowledging knowledge gaps.

- Experiment with personal AI tools to understand augmentation firsthand.
- Model the learning behaviors expected across the organization.

Phase 2: System Architecture

Next, leaders design the structures that enable swarm effectiveness:

- Establish initial swarm models with clear guidance mechanisms.
- Implement decision frameworks that balance autonomy with strategic alignment.
- Develop measurement systems that track both outcomes and behaviors.
- Launch high-visibility pilot initiatives with explicit learning goals.

Phase 3: Cultural Embedding

Finally, leaders embed transformation into the organization's DNA:

- Scale successful swarm practices across teams and functions.
- Integrate new behaviors into performance reviews and reward systems.
- Celebrate wins through storytelling that reinforces shared values.
- Continuously refine approaches based on emerging patterns and feedback.

The stakes couldn't be higher. Traditional leaders become organizational bottlenecks in AI-powered markets. Swarm orchestrators unleash collective intelligence that competitors cannot match. The choice is stark: evolve or exit.

This new leadership paradigm requires sophisticated governance frameworks to ensure ethical, sustainable transformation. Without robust guardrails, even brilliant orchestration leads to organizational catastrophe.

BUILDING TRUST THROUGH GOVERNANCE – CONVERTING ETHICS INTO COMPETITIVE MOAT

As AI capabilities raise societal concerns, governance shifts from compliance burden to strategic differentiator. Organizations that embed ethics into their operational DNA build trust advantages that technology alone cannot deliver.

Trust has become a scarce resource in the digital economy. Organizations that systematically build trust through transparent AI governance create sustainable differentiation that technology alone cannot replicate. This transforms ethical considerations from constraints into catalysts for growth.

The governance architecture for the AI era requires sophistication across four dimensions:

Strategic Foundation: Leading organizations position ethics as core business strategy rather than regulatory obligation. They recognize that trustworthy AI creates market differentiation, talent advantages, regulatory goodwill, and partnership opportunities unavailable to competitors.

Institutional Infrastructure: Effective governance requires robust organizational structures with real authority. AI Ethics Boards with P&L responsibility rather than advisory status ensure ethical considerations influence business decisions. Cross-functional governance councils break down silos that obscure risk. Real-time monitoring systems detect emerging issues before they become crises.

Operational Framework: Governance must embed directly into development cycles rather than functioning as after-the-fact review. Ethical checkpoints become integral development milestones. Automated bias detection runs alongside technical performance testing. Continuous compliance monitoring ensures systems remain aligned with values as they evolve.

Technical Architecture: The most sophisticated organizations build ethics directly into their technical infrastructure. Explainability tools make AI decision processes transparent. Privacy-preserving techniques like federated learning minimize data exposure. Monitoring systems detect model drift before it creates harm.

In the age of AI, organizations that earn trust from employees, customers, and regulators unlock measurable business benefits across brand, risk, adoption, talent, and innovation. These drive real business outcomes:

Stronger Brand Loyalty: Ethical AI leadership builds credibility and customer trust, creating a "reputation moat" that protects market position.

Lower Regulatory Risk: Transparent governance reduces the chances of fines, lawsuits, and costly compliance issues.

Faster Adoption: When employees and customers trust AI systems, implementation is smoother and usage grows more quickly.

Better Talent Attraction and Retention: A commitment to fairness and transparency boosts employee pride and makes the company more appealing to top talent.

Greater Innovation: Trust and psychological safety empower employees to propose new AI use cases and solve problems creatively.

Organizations that invest in human-centered change management consistently outperform those that don't. For example:

Higher Success Rates: AI projects with strong change management are far more likely to meet their goals than those without it.

Improved Productivity and Efficiency: Engaged employees adopt AI tools more effectively, leading to better output and smarter workflows.

Sustained ROI: Measuring both efficiency and human-centered metrics helps organizations optimize impact over time.

Reduced Costs: A human-centric approach lowers turnover and minimizes costly rework, saving time and resources.

Strategic Advantage: Organizations that master the psychology of change execute faster and more effectively, gaining a durable competitive edge.

As AI capabilities expand exponentially, trust becomes the ultimate differentiator. Organizations building systematic trust create competitive advantages that competitors cannot easily replicate. This foundation of trust enables the most ambitious transformation: using AI to enhance human fulfillment at scale.

HUMANIZING THE SWARM – SYNTHETIC SELF-ACTUALIZATION AS ULTIMATE ADVANTAGE

Beyond efficiency and innovation lies an even greater opportunity: using AI to elevate human potential itself. Organizations that design for human flourishing create sustainable advantages that extend far beyond technological implementation.

Synthetic self-actualization represents the apex of organizational evolution: intentionally designing AI systems that maximize human potential, purpose, and fulfillment. This approach transforms work from economic transaction into a vehicle for human flourishing.

This new frontier of AI leadership redefines the purpose of technology in the workplace. At its highest expression, synthetic self-actualization transforms work from a transactional necessity into a platform for personal growth and fulfillment. This shift unfolds across three perspectives:

- **Traditional view:** AI replaces human labor to reduce costs
- **Evolved view:** AI eliminates drudgery to enhance productivity
- **Revolutionary view:** AI actively cultivates human self-actualization at scale

Synthetic self-actualization represents the pinnacle of organizational design, where AI systems are intentionally built to unlock human potential, purpose, and fulfillment.

Design principles are what separate AI-era leaders from those who fall behind.

In organizations that thrive, AI is a catalyst for human empowerment, growth, and wellbeing. These leaders embed intentionality into every layer of system design, ensuring that technology amplifies what makes us uniquely human.

They begin by preserving human agency. Rather than allowing algorithms to dictate decisions, they design systems that offer intelligent recommendations while keeping final authority in human hands. For instance, an AI assistant might suggest optimal scheduling or resource allocation, but the user retains full control, ensuring augmentation without eroding autonomy.

Next, they adopt a growth orientation. Every AI interaction becomes an opportunity to build capability. Imagine a system that analyzes work patterns and proactively recommends personalized learning modules, turning routine tasks into moments of development.

They also prioritize purpose connection. Impact is made visible and tangible. A product designer, for example, might receive real-time feedback on how material choices affect sustainability goals, linking everyday decisions to broader environmental outcomes.

These leaders embrace individuation, recognizing that fulfillment is personal. A task routing system could allow employees to define interest profiles, ensuring that assignments align with what energizes them most - whether that's creative problem-solving, customer engagement, or technical depth.

Finally, they invest in holistic wellbeing. AI systems are designed to support mental, emotional, and even spiritual dimensions of work, not just productivity. A workplace analytics platform might monitor energy patterns and suggest recovery periods, helping employees maintain sustainable performance over time.

The competitive edge of human-centric AI design extends far beyond short-term productivity gains. Organizations that embrace this approach unlock deeper, more sustainable advantages across every dimension of performance.

- They become talent magnets, attracting top performers who increasingly seek purpose and fulfillment over compensation alone. In these environments, meaningful AI-augmented work becomes a differentiator in the war for talent.
- They accelerate innovation velocity, as fulfilled and engaged employees generate breakthrough ideas at significantly higher rates than their disengaged peers. When people feel valued and empowered, creativity flourishes.
- They build organizational resilience, with purpose-driven cultures that weather disruption more effectively. During times of uncertainty, employees in these organizations show greater commitment and adaptability.
- They command a brand premium, as customers increasingly choose (and are willing to pay more for) products and

services from companies that demonstrate a human-centered ethos.
- And they rise to ecosystem leadership, attracting partners who want to collaborate with organizations that reflect evolved values and shared purpose.

The future of work shaped by human-centric AI is not just an upgrade, it's a radical reimagining of employment itself.

In this emerging model, organizations evolve from being economic engines into platforms for human development. AI shifts from a perceived threat to a trusted partner in human flourishing. Business success no longer depends on extracting value from human effort, but on enabling human success as its foundation. And work itself transforms from a necessity into a primary vehicle for self-realization.

This isn't a utopian ideal; it's a pragmatic strategy for forward-looking organizations that understand the unmatched competitive advantage of fully actualized human potential. Realizing this vision requires courage: the courage to challenge long-held assumptions about work, leadership, and the relationship between technology and humanity.

SUMMARY

The AI revolution presents organizations with their most profound strategic challenge and opportunity in generations.

The AI revolution demands fundamental organizational transformation beyond technology adoption. Traditional hierarchies are giving way to fluid collaborative swarms where humans and AI systems form dynamic, adaptive networks. Organizations that recognize this shift position themselves for unprecedented advantage, while those clinging to conventional structures risk obsolescence.

These insights converge into a transformation equation that determines success:

Bold Vision + Systematic Execution + Human-Centric Design = Exponential Advantage

Key Takeaways:

- AI-human collaborative networks outperform traditional hierarchical structures in rapidly changing environments
- Systematic task decomposition creates the foundation for strategic work allocation between humans and machines
- Leadership must evolve to orchestration of collective intelligence
- Ethics and governance become competitive differentiators
- Organizations that design for human fulfillment create sustainable advantages in talent, innovation, and resilience

Chapter 2 explores the strategic frameworks for assessing your organization's current AI readiness and identifying high-impact transformation opportunities specific to your industry context and competitive landscape.

2

THE FUTURE OF WORK – A TASK-BASED ECONOMY

The real work in organizations has never lived inside job descriptions. Titles were designed to create stability, not accuracy. They tell us who reports to whom, but almost nothing about how problems actually get solved. That illusion survived when work moved slowly and systems were rigid. AI breaks it instantly.

Artificial intelligence now absorbs routine, data-heavy tasks before most employees finish their first coffee. This shift forces a fundamental question: if the "job" no longer contains the work, how should organizations think about roles? The answer emerging across industries is a task-based model where human and machine capabilities are matched dynamically to the demands of the moment.

Early adopters are already seeing the benefits. Decision cycles shrink, errors decline, and human talent is freed from repetitive work to focus on judgment, creativity, and strategy (the things that actually change outcomes). The biggest advantage isn't lower cost, but sharper allocation of human intelligence to the problems that matter most.

Yet the transition is not frictionless. When traditional jobs fragment into fluid task portfolios and your closest collaborator might be an algorithm, familiar questions surface. How do you sustain motivation when roles blur? What does career progression look like when assignments shift dynamically? And how do you maintain culture when half your team runs in the cloud?

To make sense of this shift, we need to move beyond slogans and look directly at the mechanics of the task-based model. What does it actually look like when organizations abandon static job descriptions for dynamic task allocation? How do leaders orchestrate humans and AI without losing accountability or culture? Let's break down the anatomy of this model and why it's becoming the operating system for the future of work.

WHY WORK STOPPED MAKING SENSE

The actual work happening inside companies bears little resemblance to what's written on business cards.

Work no longer follows the neat boundaries of traditional job descriptions. The shift from role-based to task-based operations has been building for decades; technology just hit the accelerator. While org charts still show tidy boxes with clear reporting lines, the reality is far messier: people operate across multiple projects, collaborate with AI systems, and adapt their responsibilities weekly.

The industrial model made sense when it was invented. Titles carried weight, specialization drove efficiency, and you could predict next year's work from this year's job description. Organizations hired for specific roles, each with defined responsibilities that changed slowly, if at all.

Technology systematically dismantled those boundaries. Assembly lines fragmented physical work into discrete motions. Computers transformed administrative tasks from human routines to automated processes. Each wave of innovation carved away another chunk of the traditional "job," replacing stable roles with dynamic task clusters.

The past decade accelerated this fragmentation. Cloud platforms made resources instantly scalable. Mobile apps put enterprise tools in everyone's pocket. Workflow automation turned complex processes into simple triggers. The gig economy normalized task-based work, while agile methodologies pushed even traditional companies toward fluid team structures.

Research confirms what practitioners already know: the skills required for the average job have shifted 25 percent since 2015 and will transform by 50 percent before 2030. So even if your title stays constant, your actual work won't. The gap between official job descriptions and daily reality keeps widening.

Organizations are at a crossroads. They can continue to rely on traditional role structures, even as work increasingly flows through informal networks and creative workarounds. Or they can embrace the evolving nature of work by designing systems that flexibly align tasks with people's strengths.

The most adaptive companies have already made the shift. They view work as a flow of tasks to be optimally distributed, not a set of roles to be preserved. This is a fundamental reimagining of how value gets created and captured in modern organizations.

WHY AI SHATTERED THE JOB DESCRIPTION

AI is coming for your to-do list, one task at a time.

Unlike humans, AI systems aren't hired for broad responsibilities. They're deployed for specific functions: analyzing datasets, recognizing patterns, answering questions, generating reports. This precision exposes how artificial our job categories have become. When machines can handle discrete tasks better than humans, the entire edifice of role-based work starts to crumble.

Consider today's project manager. While they may still update timelines and coordinate meetings, many routine tasks like status tracking and scheduling are increasingly handled by AI tools. Project managers can now focus more on navigating risks, understanding stakeholder dynamics, and steering strategic decisions. The nature of the work is evolving, with tasks flowing to whoever (whatever) is best equipped to handle them.

As specific tasks migrate to AI, human roles become fluid bundles of responsibilities that reconfigure based on real-time needs. The bank teller becomes a financial advisor. The radiologist shifts from image reading to treatment strategy. The assembly worker evolves into a quality specialist. AI strips away the repetitive core, revealing more valuable human work underneath.

Organizations embracing this shift are abandoning static job specifications for dynamic capability matching. Internal talent marketplaces let employees contribute across departments based on skills, not titles. AI systems flow between functions, handling tasks wherever efficiency gains exist. Work becomes a real-time optimization problem, not a fixed allocation exercise.

Decision-making accelerates when routine analysis happens instantly. Quality improves as AI handles precision tasks while humans focus on judgment calls. Capacity scales without proportional headcount growth. Most importantly, employee satisfaction often increases as mind-numbing tasks disappear and more engaging work emerges.

The implications run deep. Managing roles becomes insufficient; success requires orchestrating a constantly evolving task ecosystem. Competitive advantage belongs to organizations that can rapidly identify, decompose, and reallocate work based on changing conditions. The first step is a clear picture of the real work being done - not the work described on org charts, but the tasks that actually drive outcomes.

MAPPING THE WORK THAT ACTUALLY HAPPENS

AI opportunity lives in the gap between what you think your team does and what they actually do.

Moving to task-based operations requires understanding the actual work happening across your organization. This means looking beyond job titles and department labels to identify the specific, repeatable activities that consume time and create value. The exercise sounds simple. The insights it reveals are often intriguing.

Start with a comprehensive task audit. Map what people actually do. The gaps between official duties and daily reality expose both inefficiencies and opportunities. That senior analyst spending hours on data cleanup? The customer service manager manually updating five different systems? These hidden tasks represent immediate automation potential that traditional role analysis would miss.

The most valuable insights come from the people doing the work. They know which tasks eat time without adding value. They see the workarounds, the duplicate efforts across teams, the critical activities that somehow lack clear ownership. A sales manager might reveal they spend more time fighting the CRM than talking to clients. An engineer might show how they've automated personal workflows that could benefit the entire team. These frontline perspectives are gold for task redesign.

Organizations can capture this intelligence through structured approaches: workflow mapping sessions, time tracking studies, digital process mining. The key is gathering data at sufficient granularity. "Process orders" is too vague. "Extract customer data from emails," "verify shipping addresses," and "update inventory system" are the level of detail to aim for.

Even before introducing AI, this exercise delivers value. It clarifies accountability, reveals process bottlenecks, and identifies quick wins for improvement. One manufacturing company discovered three departments were independently maintaining customer contact databases. A healthcare provider found nurses spending 40 percent of their time on documentation that could be streamlined. These insights alone justify the effort.

The real power comes from systematic categorization. Not all tasks are created equal. Understanding their fundamental nature determines optimal treatment in the new operating model. Some will be automated completely. Others will be augmented. Many will be redesigned. A few will be eliminated entirely. But first, you need to see them clearly.

CATEGORIZING TASKS: BUILDING A STRATEGIC FRAMEWORK

Not every task deserves the same treatment.

Once organizations have mapped their task landscape, the next critical step involves applying a structured taxonomy that shows where AI can create maximum value while preserving essential human contributions. This classification system serves as the foundation for intelligent transformation, preventing both over-automation and missed opportunities.

Five dimensions help organizations evaluate their task portfolio:

Routine vs. Non-Routine: Routine tasks follow predictable patterns and can be documented in standard operating procedures, making them prime candidates for automation. Invoice processing, data validation, and report generation make up this category. Non-routine tasks require adaptation to novel situations, creative problem-solving, or nuanced judgment calls that current AI systems struggle to replicate effectively.

Cognitive vs. Physical: This distinction helps organizations select appropriate technologies for automation. Cognitive tasks involving information processing, analysis, and decision-making align with AI and machine learning capabilities. Physical tasks requiring manipulation of objects or navigation of real-world environments typically demand robotics solutions, though the boundary between these categories continues to blur as technology advances.

People-Facing vs. Back-Office: Tasks involving direct human interaction often require emotional intelligence, empathy, and communication skills that remain mostly human domains. Back-office operations, particularly those involving data processing or system maintenance, generally offer clearer automation pathways without sacrificing the human elements crucial to customer relationships or team dynamics.

High-Value vs. Low-Value: Value assessment must consider both immediate financial impact and strategic importance. High-value tasks directly advance organizational objectives, enhance customer outcomes, or require scarce expertise. Low-value activities, while often necessary, consume resources without proportional benefit and frequently represent the lowest-hanging fruit for automation initiatives.

Frequency and Volume: High-frequency, high-volume tasks provide compelling returns on automation investment due to scale effects. However, infrequent but high-stakes activities might also merit AI support if the consequences of error are significant enough to justify the implementation costs.

These dimensions interconnect rather than operate in isolation. A task might be routine, cognitive, back-office, low-value, and high-frequency, making it an ideal automation candidate. Conversely, a non-routine, people-facing, high-value activity would likely remain human-centered while potentially benefiting from AI augmentation tools.

THE ROI REALITY: WHICH TASKS ACTUALLY PAY OFF

Organizations waste millions automating the wrong tasks. The difference between transformative AI investment and expensive failure lies in rigorous ROI analysis.

Consider a Fortune 500 insurance company that spent $2.1 million implementing AI for complex underwriting decisions. The system achieved 73 percent accuracy - worse than their human underwriters' 89 percent rate. Meanwhile, a competing firm invested $400,000 automating routine claims documentation, reducing processing time by 78 percent and saving $3.2 million annually. Technical feasibility doesn't guarantee business value.

Effective ROI analysis examines five factors that determine whether task automation creates real value:

Complexity and Predictability: Tasks with clear rules and limited exceptions deliver immediate returns. An e-commerce company automated its customer service chatbot to handle routine tasks, aiming to free up human agents. This led to a drop in customer satisfaction when the bot failed to address complex or emotional issues. The company then improved the bot to escalate complex cases to human agents with full context, enhancing both efficiency and the customer experience.

Volume Economics: High-frequency tasks compound savings dramatically. When a telecommunications company automated password resets (a task performed 127,000 times monthly) they saved $1.4 million annually despite the modest $11 per-incident cost. Conversely, automating quarterly board presentation preparation saved only $18,000 yearly despite eliminating 60 hours of executive assistant time.

Error Costs and Compliance Risk: In regulated industries, accuracy improvements alone can justify automation. A pharmaceutical company reduced FDA compliance violations by 94 percent after implementing AI for adverse event reporting, avoiding potential fines exceeding $10 million. Their competitor's rushed automation of clinical trial documentation, however, introduced errors that delayed drug approval by eight months, costing an estimated $320 million in lost revenue.

Integration Complexity: Standalone tasks automate more successfully than deeply interconnected workflows. A retailer achieved 91 percent automation of inventory reordering by focusing on a self-contained process. Their attempt to automate the entire supply chain workflow failed after 18 months and $4.7 million investment, primarily due to system integration challenges across 14 different platforms.

Human Transition Costs: Successful automation accounts for workforce adaptation. A global firm budgeted 15 percent of their AI initiative for employee retraining, successfully transitioning 89 percent of affected staff to higher-value roles. Another professional services firm ignored this factor, leading to 31 percent turnover among displaced workers and $2.3 million in recruitment and knowledge transfer costs.

The most sophisticated organizations now use weighted scoring models combining these factors. Tasks scoring above predetermined thresholds proceed to pilot programs, while others undergo redesign or remain human-executed. This disciplined approach separates the 5 percent of AI initiatives that deliver substantial ROI from the 95 percent of generative AI pilots that fail, according to published reports.

DESIGNING HUMAN-AI TEAMS THAT ACTUALLY WORK

The most expensive AI implementations fail not because of technology limitations, but because organizations design them like traditional IT projects instead of human-machine partnerships.

A published analysis of 1,200 AI implementations found that projects emphasizing human-AI collaboration achieved 3.5x better outcomes than those focused purely on automation. The difference? Successful organizations design for synergy, not substitution.

Clear Handoff Protocols Prevent Costly Confusion

In a study on cancer diagnostics, the automated diagnostic method achieved 92 percent accuracy, while human pathologists achieved 96 percent. When the two were combined in a "human-in-the-loop" model, the accuracy improved to 99.5 percent, significantly reducing errors.

Transparency Builds Trust and Adoption: A major teaching hospital introduced AI diagnostic support by showing radiologists exactly which image features triggered specific recommendations. Adoption reached 94 percent within six months, compared to 31 percent at a peer institution that deployed a "black box" system. When clinicians understand AI reasoning, they integrate it into their workflow rather than working around it.

Feedback Loops Enable Continuous Improvement: A direct to consumer firm's recommendation engine constantly improves because users actively correct its mistakes. Similarly, a manufacturing company that implemented AI quality control saw defect detection accuracy increase from 82 percent to 97 percent over 18 months by systematically incorporating inspector feedback.

Hybrid Workflows Outperform Pure Automation: According to published reports, Google's internal research found that human-AI teams consistently outperform either humans or AI alone in complex tasks. Their code review process combines automated scanning with human expertise: AI catches syntax errors and security vulnerabilities while humans evaluate architecture and maintainability. Development velocity increased 35 percent with 50 percent fewer production bugs.

Task Orchestration Requires New Management Skills

Leading organizations are creating "swarm conductor" roles: managers who optimize task distribution between humans and AI in real time. A global logistics company reduced delivery times by 23 percent by dynamically routing urgent shipments to human dispatchers while AI handled routine scheduling.

Organizations that view AI as a team member rather than a replacement achieve superior results. This requires designing systems where machine precision complements human judgment, where automation handles scale while people manage exceptions, and where both sides of the partnership continuously learn from each other.

REBUILDING JOBS FOR THE TASK ECONOMY

Static role definitions can't capture work that shifts based on AI capabilities and business demands.

Leading organizations are abandoning rigid job specifications for dynamic frameworks that emphasize outcomes over activities. Instead of listing 15 required tasks, modern role descriptions focus on problems to solve and value to deliver. A product manager's mandate becomes "accelerate feature delivery using available human and AI resources" rather than "conduct daily standups and maintain project documentation."

This shift demands new hiring practices. Skills-based recruitment replaces credential checking. Internal talent marketplaces let employees flow between projects based on capability match rather than departmental assignment. A financial firm's internal platform handles tens of thousands task-based assignments annually, reducing project staffing time from weeks to hours while increasing employee satisfaction through greater variety and growth opportunities.

The Emergence of Fusion Roles

New positions blur traditional boundaries between human and machine work. The "AI whisperer" helps marketing teams craft prompts that generate compelling content. The "automation architect" identifies workflow bottlenecks across departments and designs human-AI solutions. The "algorithmic auditor" ensures AI systems align with company values and regulatory requirements.

These roles demand hybrid skillsets: technical fluency without requiring engineering depth, business acumen combined with data literacy, change management expertise paired with system thinking. The 'fusion worker' sounds like something developed in a nuclear lab, but it's actually just an employee who can code a little, write a little, and not panic when the AI starts hallucinating. A financial services firm found their best fusion workers were philosophy majors, proving that after 2,000 years, a degree in Philosophy finally has a positive ROI. Who better to debate the ethics of a neural network than someone who spent four years deconstructing Kant?

Career Paths Become Career Webs

Linear progression gives way to multidirectional growth. Employees might spend two years optimizing supply chain algorithms, pivot to customer experience design, then leverage both experiences to lead digital transformation initiatives. This lattice model rewards adaptability and cross-functional expertise over deep specialization in narrowing fields.

Performance evaluation frameworks must evolve accordingly. Annual reviews based on static job descriptions become obsolete when roles shift quarterly. Progressive organizations implement continuous feedback systems that assess project outcomes, skill development, and adaptation capability. What matters isn't whether someone fulfilled predetermined duties, but whether they delivered value while growing their capabilities.

Preparing for Perpetual Evolution

Smart companies now include "role evolution expectations" in employment discussions. They explicitly acknowledge that positions will transform as AI capabilities expand. Rather than creating anxiety, this transparency often attracts talent seeking dynamic environments over static security.

One global consulting firm redesigned their entire organizational structure around "impact teams" - fluid groups that form around client challenges, combining human experts with AI tools as needed. Traditional departments still exist for administrative purposes, but actual work happens in these adaptive units. Employee engagement scores increased significantly as people gained exposure to diverse projects and developed broader skillsets.

The most forward-thinking organizations view job descriptions as living documents, updated quarterly based on technological capabilities and business needs. They're building workforces comfortable with ambiguity, skilled at human-AI collaboration, and energized by continuous learning rather than threatened by change.

PREPARING THE WORKFORCE WITHOUT BREAKING TRUST

Most AI transformations fail because organizations focus on technology deployment while ignoring the humans who must work alongside it. Smart leaders prepare their people first.

Successful workforce preparation starts with honest communication about what's changing and why. When a major software firm introduced AI coding assistants, they led with transparency: "This will change how you write code, not whether you have a job writing it." They backed this message with comprehensive reskilling programs that helped 89 percent of developers transition to higher-level design and architecture work within six months.

Address the Fear Before It Festers

Anxiety about AI replacement is rational, although it can look like resistance. Acknowledge it directly. A European manufacturer facing automation of assembly line tasks created "future role workshops" where employees explored how their jobs would evolve. Workers who initially opposed the changes became advocates after seeing concrete pathways to new positions in quality control, robot maintenance, and process optimization.

The most effective approach combines individual career coaching with team-based transition planning. When a major retailer automated inventory management, they assigned each affected employee a "transition partner": a colleague already working in their target role. This peer mentorship model achieved 25 percent more successful role transitions compared to using traditional training alone.

Build New Capabilities Through Real Work

Create learning opportunities within daily operations. A healthcare system introducing AI diagnostic tools embedded training directly into clinical workflows. Doctors learned by reviewing AI suggestions during real patient encounters, with expert support immediately available. Competency development accelerated while patient care quality improved.

Skills development must span technical and human domains equally. As AI handles routine analysis, professionals need stronger abilities in:

- Interpreting AI outputs with business context

- Managing human-AI team dynamics
- Identifying when to override automated recommendations
- Communicating complex technical concepts to stakeholders
- Maintaining ethical oversight of AI systems

Make Change Champions, Not Change Victims

Involve employees in designing their AI-augmented future. An insurance company created "automation advisory boards" staffed by frontline workers who identified which tasks to automate first. These employees became natural evangelists, helping colleagues see AI as a tool they controlled rather than a threat they endured.

Celebrate adaptation publicly. When a logistics company highlighted employees who successfully transitioned from route planning to customer relationship management, applications for reskilling programs jumped 200 percent. Success stories from peers carry more weight than executive assurances.

Support the Whole Person

Career transitions create stress beyond the workplace. Progressive organizations offer comprehensive support: financial planning assistance for those whose roles are changing, family counseling to address home anxieties, and extended healthcare coverage during transition periods. This holistic approach reduces turnover and builds loyalty that extends beyond the transformation period.

One technology company discovered that employees' biggest concern wasn't job loss but becoming irrelevant. They created "relevance pathways", personalized development plans showing how current skills could evolve into future capabilities. Engagement scores increased as workers saw clear connections between who they were and who they could become.

The organizations succeeding in AI transformation share a common thread: they invest as much in their people as in their technology. They recognize that sustainable change happens at human speed, not digital speed. Most importantly, they understand that workforce preparation isn't a phase of AI implementation, it's the foundation upon which everything else is built.

THE PATH FORWARD: BUILDING ORGANIZATIONS THAT LEARN

The future belongs to companies that treat transformation as a permanent state, not a one-time project. Task-based work requires organizations that continuously adapt their structures, processes, and capabilities.

The most successful AI adopters share three critical characteristics that distinguish them from companies still struggling with basic automation. First, they've abandoned the fiction of stable end states. Their operating models assume constant evolution as AI capabilities advance and business conditions shift. Second, they've built learning directly into their organizational DNA through systematic experimentation and rapid iteration. Finally, they measure success through adaptability metrics alongside traditional performance indicators.

Consider how leading organizations approach continuous transformation:

Experimentation Becomes Systematic: A leading business to consumer firm's "working backwards" methodology now includes AI impact assessment for every new initiative. Teams must identify which tasks could be automated, which require human judgment, and how the human-AI mix might evolve over 18 months. This forward-looking discipline prevents creating roles or processes that become obsolete before full implementation.

Measurement Evolves Beyond Efficiency: Traditional metrics miss the full impact of task-based transformation. Progressive organizations track:

- **Capability velocity:** How quickly can the organization deploy new AI tools and retrain workers?
- **Task fluidity:** What percentage of work flows seamlessly between human and AI execution?
- **Innovation density:** How many new value-creating activities emerge as routine tasks automate?
- **Adaptation index:** How effectively do employees navigate role transitions and skill development?

A global pharmaceutical company discovered their most valuable metric was "time to first patient benefit": how quickly AI-assisted drug discovery translated into clinical trials. This measure aligned their entire organization around meaningful outcomes rather than process efficiency.

Culture Becomes the Sustainable Advantage

Technology advantages erode quickly as AI tools commoditize. Organizational culture (the shared beliefs, behaviors, and assumptions about how work happens) provides durable differentiation. Companies with cultures embracing experimentation, continuous learning, and human-AI collaboration consistently outperform those clinging to traditional hierarchies and fixed roles.

In a global financial services firm, a strategy team deployed a generative AI model to synthesize macroeconomic reports, regulatory filings, and earnings statements. The idea was to create a faster forecast model that acted as a real-time research assistant. What emerged was unexpected. A mid-level analyst began feeding the same AI outputs into a second tool to generate "What if" scenarios, e.g., "What if interest rates rise by 1.25 percent and consumer confidence drops below 40?" These prompts weren't in the brief. Soon, the analyst began tracking how multiple teams responded to these outputs. The AI had provided information, but the human had abstracted it into action triggers. The OCM team formalized the analyst's approach into a Scenario Planning Playbook, showing how to use AI outputs to surface options. A templated abstraction framework was developed and used in quarterly planning across four business units. What started as content acceleration turned into a cultural shift: from insight consumption to insight transformation.

Leadership Transforms from Directors to Orchestrators

The executive role fundamentally changes in task-based organizations. Rather than managing through hierarchy and control, leaders become orchestrators who:

- Set vision and connect AI outcomes to the organizational priorities while allowing teams to determine optimal human-AI task distribution
- Remove obstacles that prevent fluid collaboration between people and machines
- Ensure ethical AI deployment while maintaining business velocity
- Model continuous learning by publicly sharing their own skill development journeys

One CEO of a Fortune 500 retailer now spends Friday afternoons working alongside frontline employees using new AI tools. This hands-on approach signals that adaptation happens at every level while providing unfiltered insight into implementation challenges. What starts as leadership modeling on the front lines must scale into an organization-wide system for sensing, sharing, and reusing what AI surfaces, hence the need to curate networks that rapidly capture and redistribute knowledge - knowledge arbitrage networks.

Curating Knowledge Arbitrage Networks

In AI-native environments, knowledge is overwhelming. The true differentiator is how fast organizations can make sense of what AI produces, redistribute that insight across boundaries, and remix it into new value. That changes the mandate for organizational change management (OCM).

Change leaders must now act as curators of knowledge arbitrage networks, spotting where valuable human-AI insight is generated and ensuring it circulates through the organization. That includes:

- Tapping into informal influencers and edge experimenters.
- Connecting siloed teams working on overlapping problems.
- Facilitating knowledge reuse across projects, not just behavior alignment within them.

OCM also plays a central role in supporting decentralized knowledge generation. As generative AI tools land in the hands of frontline workers, change agents must build the conditions where experimentation can flourish without waiting for top-down approval. Instead of delivering pre-scripted best practices, OCM enables adaptive loops of insight, where learning emerges organically and is rapidly socialized.

Finally, the work of OCM shifts from "resistance mitigation" to sense amplification. The challenge is helping employees navigate AI noise, make meaning in uncertainty, and learn how to think with machines. This is less about controlling the message, and more about increasing cognitive clarity at scale. Some ways to put this into practice:

Weekly AI check-in: Spend 10 minutes at the end of the week sharing one useful AI insight, the evidence behind it, and what action it triggered.

Insight library: Keep a shared document or board where teams post short summaries of their best AI findings so others can reuse them.

Teach the basics: Offer quick tips on how to read AI confidence scores, check sources, and write better prompts.

The Integration Imperative

None of these elements work in isolation. Systematic experimentation requires cultural permission to fail. New metrics demand leadership commitment to measure what matters. Cultural transformation needs structural support through revised roles and workflows. The organizations winning in the AI era integrate all these components into a coherent operating system designed for perpetual evolution.

The path forward is about building the capability to thrive amid continuous change. Organizations that master this meta-skill create compounding advantages: they implement new technologies faster, adapt to market shifts more effectively, and attract talent energized by dynamic environments.

The task-based economy rewards organizations that embrace impermanence. Success belongs to those who build systems, cultures, and capabilities optimized for constant transformation. Everything else becomes legacy overhead in a world where change is the only constant.

SUMMARY

In this chapter, we explored the seismic shift from a role-based to a task-based economy. We dismantled the illusion of the static job description, revealing how AI exposes the gap between official titles and the actual, fluid nature of work. By mapping workflows at the granular level, organizations can identify where human judgment creates the most value and where machine precision offers the greatest efficiency. This is not merely an exercise in automation, but a strategy for liberation, freeing talent from drudgery to focus on fusion roles that blend technical fluency with creative problem-solving. The future belongs to organizations that treat work as a dynamic portfolio of capabilities rather than a rigid hierarchy of positions.

Chapter takeaways:

- Break traditional roles into specific tasks; let AI handle repetitive, data-heavy work so humans can focus on judgment, creativity, and relationship-building.
- Map out what employees really do day-to-day to uncover hidden inefficiencies and automation opportunities that org charts miss.
- Evaluate tasks by routine vs. non-routine, cognitive vs. physical, people-facing vs. back-office, value, and frequency to decide what to automate, augment, or redesign.
- Use ROI analysis to automate high-volume, low-complexity, or error-prone tasks, but keep humans in the loop for complex or high-risk activities.
- Build workflows where AI and people complement each other, with clear handoffs, transparent AI reasoning, and feedback loops for mutual improvement.
- Move to dynamic, outcome-focused roles and encourage hybrid "fusion" jobs that blend technical, business, and change skills.
- Invest in ongoing learning so employees can move into higher-value roles as AI capabilities expand.
- Redesign work so automation enriches jobs, not just eliminates tasks; keep employees involved in shaping their AI-augmented future.
- Create mechanisms for employees to give feedback on AI, and for AI to learn from human corrections, driving continuous improvement.
- Treat transformation as ongoing: embed experimentation, rapid iteration, and learning agility into your organization's DNA.

With the nature of work fundamentally reimagined, the demands on those at the helm have never been greater. A flexible, AI-integrated workforce cannot be managed by a rigid, industrial-era hierarchy. This brings us to the next critical frontier: leadership. How do executives move from directing employees to orchestrating collective intelligence? Chapter 3 explores this pivotal evolution, outlining the strategic vision, emotional intelligence, and adaptive frameworks required to lead organizational transformation in the AI era.

3

LEADING ORGANIZATIONAL TRANSFORMATION IN THE AI ERA

The AI revolution demands more than technical expertise; it requires adaptive leadership capable of orchestrating human and machine intelligence.

Organizations aren't just adopting new technologies; they're fundamentally reimagining how work gets done, decisions are made, and value is created. For senior leaders, the question isn't whether to embrace AI but how to guide their organizations through transformation while preserving their human core.

Artificial intelligence weaves through every industry with remarkable velocity. Healthcare systems deploy predictive algorithms that anticipate patient deterioration before clinicians can detect warning signs. Financial institutions harness machine learning to identify fraudulent transactions in milliseconds, preventing billions in losses annually. Manufacturing plants orchestrate robot-human teams that achieve unprecedented precision and safety records. Yet beneath these technical triumphs lies a more complex challenge: the human dimension of change.

Consider the paradox facing today's executives. They must champion AI adoption to remain competitive while managing the anxiety it creates among their workforce. They need to drive efficiency gains through automation while fostering the creativity and problem-solving capabilities that only humans possess. Most critically, they must navigate this duality without compromising organizational values or losing the trust of employees, customers, and stakeholders.

Recent studies reveal a sobering truth: only 5 percent of AI pilots succeed. The culprit isn't flawed algorithms or insufficient computing power; it's inadequate attention to organizational factors. Resistance to change, strategic misalignment, and communication breakdowns derail more AI initiatives than technical glitches.

The most successful AI transformations share common characteristics. Leaders articulate compelling visions that connect AI capabilities to strategic imperatives. They develop new competencies (from data fluency to emotional intelligence) that enable them to bridge technical and human domains. They communicate with clarity and empathy, transforming fear into engagement. Perhaps most importantly, they recognize that AI adoption isn't a destination but a journey requiring continuous adaptation.

This chapter examines seven critical dimensions of AI-era leadership. (1) We begin with vision and strategic alignment, exploring how leaders establish clear direction for AI integration. (2) We then examine the evolved skill sets required to lead human-AI teams effectively. (3) Communication is the connective tissue binding organizations together during change, followed by (4) culture-building that emphasizes adaptability and resilience. (5) Ethical considerations demand proactive attention, while (6) measurement systems ensure initiatives deliver tangible value. Finally, we address the imperative of (7) sustaining momentum through continuous innovation and learning.

Throughout, we'll encounter real-world examples from healthcare, finance, and technology sectors as illustrations of principles in action.

The goal is to equip you with frameworks for navigating your unique transformation journeys. By chapter's end, you'll understand how to lead an AI transformation that amplifies human potential.

VISION AND STRATEGIC ALIGNMENT FOR AI INTEGRATION

Strategic clarity transforms AI from a collection of experiments into a coherent force for organizational advancement. The most successful AI transformations begin in boardrooms where leaders articulate why artificial intelligence matters to their organization's future. This vision serves as both compass and catalyst, guiding investment decisions while inspiring collective action across the enterprise.

Consider how leading organizations frame their AI ambitions. A global pharmaceutical company doesn't simply announce plans to "leverage AI"; it commits to "accelerating drug discovery by 40 percent while reducing development costs by $100 million annually through AI-powered molecular modeling." A retail giant moves beyond generic automation goals to declare its intention to "create personalized shopping experiences for 50 million customers while eliminating 90 percent of inventory waste through predictive analytics." These visions connect technological capability to measurable business outcomes that stakeholders can understand and support.

Strategic alignment requires more than aspirational statements. Leaders must map AI initiatives to specific organizational objectives, creating clear line-of-sight between technological investments and business value. When a financial services firm deploys machine learning for credit decisions, it aligns with strategic goals around financial inclusion and risk management. When a hospital system implements AI-powered diagnostic tools, it advances objectives around patient outcomes and clinical efficiency. Without this alignment, AI projects drift into technological curiosities that consume resources without delivering impact.

When C-suite leaders visibly champion AI initiatives (allocating budgets, removing obstacles, and demanding cross-functional collaboration) they signal that transformation is an enterprise priority, not merely an IT project. This top-level commitment becomes especially crucial when integrating AI across departments with different cultures, metrics, and incentives. A unified vision prevents the common pitfall of competing AI efforts that duplicate work or undermine each other.

Values alignment adds another critical dimension. Organizations renowned for customer privacy must ensure their AI strategies reflect that commitment through transparent data practices and explainable algorithms. Companies that pride themselves on employee development should frame AI as an augmentation tool that enhances human capabilities rather than replaces workers. When AI initiatives resonate with core organizational values, they generate authentic buy-in rather than compliance-driven adoption.

Traditional five-year roadmaps aren't sufficient in an environment where AI capabilities evolve monthly. Leading organizations adopt more agile approaches, establishing directional strategies with quarterly reviews and adjustment mechanisms. They create AI steering committees that bring together business leaders, technologists, and domain experts to continuously evaluate opportunities and risks. This governance structure ensures strategic alignment remains dynamic rather than static.

Start by identifying your organization's "AI North Star": the transformative outcome that would fundamentally strengthen your competitive position. Then work backward to define the capabilities, data assets, and cultural changes required to reach that destination. Plot initiatives along a maturity curve, beginning with high-impact, low-risk projects that build confidence and capability before tackling more ambitious transformations.

While financial returns matter, AI's strategic value often manifests in enhanced decision-making speed, improved customer experiences, or increased organizational agility - benefits that resist simple quantification. Progressive organizations develop balanced scorecards that capture both tangible outcomes (cost savings, revenue growth) and strategic indicators (innovation velocity, market responsiveness).

The vision-setting process culminates in what one technology executive calls the "believable bold": an AI strategy ambitious enough to inspire yet grounded enough to execute. This balance prevents both the timidity that wastes AI's potential and the overreach that invites failure. When leaders achieve this equilibrium, they create momentum that carries organizations through the inevitable challenges of transformation. This strategic clarity sets the stage for the next ingredient of AI transformation: leadership. With a compelling vision in place, organizations must now cultivate the human capabilities that will bring it to life.

LEADERSHIP SKILLS FOR THE AI-ENHANCED WORKPLACE

Today's executives must navigate between algorithmic precision and human intuition, orchestrating teams where artificial and human intelligence collaborate rather than compete.

Leading in the AI era requires navigating the 'Executive Pretense Paradox.' It's that special kind of imposter syndrome where you find yourself using words like 'generative,' 'modalities,' and 'inference' in sentences, hoping they are in the right order. Acknowledgement is the first step to recovery. You don't need to know how to build the watch; you just need to know how to stop the watch from hallucinating and insulting your biggest client. The learning curve is steep, but acknowledging one's knowledge gaps is the first step toward technical literacy. A healthcare CEO who grasps how machine learning algorithms learn from training data can better assess whether an AI diagnostic tool might perpetuate historical biases.

The AI-enhanced workplace generates unprecedented volumes of insights, predictions, and pattern recognition. Effective leaders develop "algorithmic appreciation" - the ability to synthesize AI-generated insights with contextual knowledge and ethical considerations. They know when to trust the data and when to override it. During the pandemic, several airline CEOs famously rejected AI models predicting quick travel recovery, relying instead on their understanding of behavioral psychology and public health dynamics. Their human judgment proved more accurate than the algorithms. This is a critical feature of a humanized swarm: The Veto Power of Intuition. While AI excels at processing historical data, humans excel at sensing shifts in the zeitgeist. A humanized organization explicitly grants leaders the authority to override the swarm when the data contradicts human values or common sense.

Paradoxically, as organizations become more automated, human-centric leadership skills grow more critical. Emotional intelligence becomes essential for guiding teams through AI-driven change. When a manufacturing supervisor introduces collaborative robots to an assembly line, workers' initial reactions often include fear, skepticism, or resentment. Leaders with high emotional intelligence can acknowledge these concerns authentically while painting a compelling picture of human-robot partnership that enhances both productivity and job satisfaction.

AI capabilities evolve at exponential rates, rendering yesterday's best practices obsolete. Leaders must model continuous learning, openly acknowledging when their assumptions need updating. One Fortune 500 CTO instituted "reverse mentoring," pairing senior executives with junior data scientists to accelerate leadership's AI fluency. This vulnerability (admitting what they don't know) paradoxically strengthened these leaders' credibility with technically sophisticated teams.

Leaders must also encompass the role of Knowledge Flow Architect. They must actively dismantle hierarchies and barriers that slow down how knowledge moves. They must cultivate internal ecosystems where ideas move horizontally, across teams, functions, and levels. This requires rotating teams, fluid decision rights, and creating visibility for bottom-up insight.

Finally, leaders must protect and promote permissionless innovation zones: environments where experimentation does not require prior approval or senior sponsorship. This means establishing trust-based parameters for safe failure, ensuring learning loops are in place, and amplifying internal wins even when they originate outside formal programs.

Every AI deployment involves value judgments, which data to use, which outcomes to optimize, which stakeholders to prioritize. Leaders must navigate these choices with moral clarity, balancing efficiency with fairness, innovation with responsibility. When a major bank's AI lending algorithm showed disparate impact across demographic groups, leadership faced a decision. Rather than tweaking the model to pass regulatory scrutiny, they fundamentally reconsidered their approach, eventually developing a more inclusive system that expanded credit access while maintaining risk standards.

AI excels at pattern recognition within defined parameters but struggles with novel situations requiring creative leaps. Leaders who can move fluidly between systematic analysis and intuitive insight create powerful human-AI partnerships. They might use AI to surface hidden patterns in market data, then apply human creativity to envision entirely new business models those patterns suggest.

Successful AI-era leaders develop what psychologists term "metacognition": awareness of their own thinking processes. They recognize when they're falling prey to automation bias (over-trusting AI recommendations) or algorithm aversion (dismissing valid AI insights due to discomfort with machine judgment). This self-awareness allows them to model balanced human-AI collaboration for their teams.

The cumulative effect of these evolved capabilities? Leaders become bridges between human and artificial intelligence, translating strategic vision into technical requirements while ensuring technological capabilities serve human values. They create environments where both people and algorithms can perform at their best, recognizing that the future belongs not to humans or machines, but to human-machine teams guided by thoughtful leadership.

To lead effectively in this new paradigm, it's not enough to simply understand AI; leaders must also guide others through its implications. That guidance hinges on one of the most timeless and critical leadership skills: communication.

COMMUNICATING EFFECTIVELY DURING AI TRANSITION

Clear communication transforms AI anxiety into engagement, converting potential resistance into active participation. Leaders must address deep-seated fears about job displacement, navigate skepticism about AI's reliability, and build confidence in radically new ways of working. The difference between successful and failed AI implementations often traces back to how effectively leaders communicate throughout the journey.

Forward-thinking leaders begin conversations months before implementation, creating space for dialogue rather than dictate. A major insurance company planning to introduce AI-powered claim processing began with town halls where the CEO personally addressed automation's impact on jobs. By acknowledging concerns upfront and outlining reskilling programs, leadership transformed potential adversaries into change advocates.

Technical jargon alienates non-technical stakeholders, while oversimplification insults intelligence. When crafting your AI narrative, avoid the trap of 'corporate glossolalia' - speaking in tongues made entirely of buzzwords. If your vision statement reads, 'We are leveraging synergistic AI modalities to paradigm-shift our cross-functional ideation,' you haven't inspired your team; you've just triggered their fight-or-flight response. Your employees don't want to hear about 'synergy'; they want to know if the robot is going to take their parking spot. Keep it simple. Effective leaders find the middle ground, using accessible metaphors without dumbing down concepts. When explaining machine learning to customer service representatives, one executive compared it to "teaching a very dedicated student who never forgets patterns but needs help understanding context." This framing made the technology approachable while accurately conveying both capabilities and limitations.

Transparency builds trust even when the news isn't entirely positive. Leaders who candidly discuss AI's potential impact on roles and workflows earn credibility for subsequent assurances about job security or growth opportunities. A manufacturing firm implementing predictive maintenance AI openly shared that routine inspection tasks would diminish but emphasized how technicians would evolve into "machine teaching" roles, training AI systems through their expertise. This honesty, paired with concrete transition plans, reduced anxiety and increased engagement.

Town halls, focus groups, and digital forums where employees can voice concerns and share ideas create ownership in the transformation process. When a healthcare system introduced AI-assisted diagnosis tools, they established physician advisory panels that shaped implementation approaches. Doctors who initially resisted "algorithmic medicine" became champions after their clinical insights influenced system design.

Regular updates (even when progress is slow) maintain momentum and prevent information vacuums that rumors fill. One financial services firm created a weekly "AI transformation digest" that shared wins, acknowledged setbacks, and celebrated employees who embraced new ways of working. This rhythm of communication normalized change as an ongoing journey rather than a disruptive event.

Board members need to understand strategic rationale and ROI projections. Middle managers require clarity on changing team dynamics and performance metrics. Front-line employees want specifics on how daily work will evolve. Successful communicators craft messages that resonate with each audience while reinforcing common themes about organizational direction and values.

Visual communication often explains AI concepts more effectively than words alone. Process diagrams showing how AI augments human decision-making, before-and-after workflow illustrations, and simple demonstrations of AI tools in action help stakeholders grasp abstract concepts. When a logistics company introduced route optimization AI, they created animated visualizations showing how the system considered multiple variables simultaneously; something impossible to convey through verbal explanation alone.

Success stories, strategically shared, build confidence and momentum. But credibility demands balanced reporting that acknowledges challenges alongside achievements. When a retail chain's AI inventory system initially increased stockouts due to incomplete training data, leadership communicated both the problem and the rapid learning curve that led to 30 percent reduction in waste within six months. This transparency strengthened trust in leadership's commitment to honest communication.

Cultural sensitivity shapes effective AI communication, particularly in global organizations. What resonates in Silicon Valley may alienate manufacturing workers in the Midwest or trigger privacy concerns in Europe. One multinational developed region-specific communication strategies that respected local values while advancing global AI objectives. In privacy-conscious markets, they emphasized data protection measures; in efficiency-focused cultures, they highlighted productivity gains.

When employees see colleagues successfully working with AI tools, skepticism diminishes faster than through top-down messaging. Progressive organizations identify and cultivate "AI ambassadors": early adopters who can authentically share experiences and coach hesitant colleagues. These peer advocates prove especially effective at translating technical capabilities into practical benefits.

When employees raise concerns about AI system limitations or suggest improvements, visible action on that feedback reinforces engagement. A healthcare system modified its AI scheduling algorithm after nurses identified patterns the system missed, then celebrated this human-AI collaboration as a win for their vision for technology partnership.

Effective AI communication ultimately creates a shared narrative about technological transformation, one where humans remain central to the organization's future. When leaders achieve this communication mastery, AI implementation shifts from something done to employees to something done with them, dramatically increasing the odds of sustainable success.

But even the most skillful communication must be reinforced by something deeper: a culture that embraces change. Without a supportive cultural foundation, even well-articulated AI strategies can falter.

BUILDING A CULTURE OF ADAPTABILITY AND RESILIENCE

Culture determines whether AI augments or threatens an organization's human potential. Consider two insurance companies deploying similar underwriting algorithms. Company A maintains a hierarchical, risk-averse culture where deviations from established process require multiple approvals. When their AI system suggests novel risk assessment approaches, employees default to traditional methods, effectively neutralizing the technology's potential. Company B cultivates experimentation and learning from failure. Their underwriters actively test AI recommendations, document outcomes, and refine the system based on real-world results. Within a year, Company B achieves 40 percent faster processing times and more accurate risk pricing, while Company A struggles to justify their AI investment.

When executives openly acknowledge their own AI learning curve (perhaps admitting confusion about certain technical concepts or sharing how they've adapted their decision-making process) they signal that growth and adaptation are organizational expectations, not signs of weakness. One pharmaceutical CEO instituted "Learning Fridays" where she publicly practiced using new AI tools alongside employees, making mistakes and asking questions without embarrassment. This vulnerability transformed organizational attitudes toward continuous learning. When a healthcare AI system misdiagnosed a rare condition, the radiologist who caught the error felt safe reporting it immediately. This transparency triggered system improvements that prevented future errors. In organizations lacking psychological safety, such mistakes go unreported, AI systems fail to improve, and potentially harmful patterns perpetuate.

Organizations claiming to value innovation while exclusively rewarding short-term efficiency send mixed signals that typically default to risk aversion. Progressive companies restructure incentives to celebrate learning achievements: documenting failed experiments that yield valuable insights, sharing knowledge across departments, or developing creative applications for AI tools. One technology firm includes "innovation credits" in performance reviews, recognizing employees who thoughtfully experiment with new approaches even when initial results disappoint.

When a retail chain's AI demand forecasting system badly miscalculated seasonal patterns, causing significant overstock, leadership resisted the urge to abandon the technology. Instead, they conducted a transparent post-mortem, identified training data gaps, and involved store managers in system refinement. Organizations often claim to celebrate failure until someone actually fails. The difference between companies that genuinely learn from setbacks and those that merely pay lip service to the concept can usually be measured by the speed with which people volunteer to lead the next high-risk AI initiative. The revised system ultimately outperformed human forecasting, but more importantly, the organization developed muscle memory for constructive response to setbacks.

Cross-functional collaboration becomes essential as AI blurs traditional departmental boundaries. Marketing's customer segmentation AI needs integration with sales' CRM system and customer service's interaction data. Organizations fostering territorial silos find their AI initiatives hamstrung by data hoarding and competing priorities. Those promoting collaborative culture (through joint objectives, shared metrics, and regular cross-pollination of ideas) unlock AI's potential for holistic optimization.

Quarterly earnings pressure can discourage the patient experimentation AI development requires. Some companies successfully building adaptive cultures create protected spaces for longer-term innovation, forming dedicated teams with multi-year mandates to explore AI applications without immediate ROI demands. These "innovation greenhouses" generate learnings that eventually benefit the entire organization.

Cultural transformation requires persistent reinforcement through multiple channels. Town halls celebrating adaptation success stories, internal newsletters highlighting creative AI applications, peer recognition programs for knowledge sharing; these seemingly small interventions compound into significant cultural shifts. One manufacturing company created an "AI Innovation Wall" where employees post creative uses they've discovered for AI tools, sparking friendly competition and viral spread of best practices.

Middle management often determines whether cultural change penetrates beyond executive rhetoric. When supervisors feel threatened by AI's potential to flatten hierarchies or expose performance gaps, they can subtly undermine adoption through skeptical comments or bureaucratic obstacles. Successful organizations invest heavily in middle management development, helping these crucial culture carriers understand how AI enhances rather than diminishes their roles.

Teams with diverse backgrounds (technical and non-technical, different industries, various cultural contexts) identify blind spots and creative applications that homogeneous groups miss. One financial services firm deliberately staffed their AI implementation team with former teachers, musicians, and social workers alongside data scientists, yielding unexpectedly innovative approaches to customer service automation.

Collaborations with startups, universities, or innovation labs expose employees to different working styles and cutting-edge thinking. These connections challenge internal assumptions and demonstrate alternative approaches to AI integration. Regular "innovation exchanges" where employees spend time in different organizational contexts often return with fresh perspectives that catalyze cultural growth.

The ultimate measure of cultural adaptability lies not in smooth AI implementation but in how organizations respond when things go wrong. Resilient cultures treat setbacks as data points. They ask "What can we learn?" rather than "Who can we blame?" This orientation transforms every AI initiative - successful or not - into a stepping stone toward greater organizational capability.

When adaptability and resilience become embedded in organizational DNA, AI integration shifts from a threatening disruption to an exciting evolution. Employees approach new technologies with curiosity rather than fear, confident in their ability to learn and adapt. As organizations grow more comfortable with continuous transformation, they must also become more reflective about the direction of that change. The next challenge is how to implement AI responsibly.

NAVIGATING ETHICAL CHALLENGES AND RESPONSIBILITIES

Every algorithm embeds human values in code, whether deliberately chosen or unconsciously inherited from biased data. Leaders who treat ethics as an afterthought risk creating systems that scale discrimination, erode privacy, or undermine human agency.

The stakes of AI ethics extend far beyond regulatory compliance. When a major retailer's pricing algorithm inadvertently charged higher prices in lower-income neighborhoods, the reputational damage far exceeded any legal penalties. Customers felt betrayed, employees questioned company values, and investors worried about governance. Conversely, when a healthcare system discovered their AI diagnostic tool performed worse for certain ethnic groups, their transparent response (halting deployment, investigating causes, and rebuilding with more diverse training data) actually strengthened stakeholder trust.

Bias represents perhaps the most pervasive ethical challenge. AI systems learn from historical data that often reflects past inequities. A technology company's resume-screening AI began filtering out women candidates because it learned from a decade of hiring data dominated by male employees. The technical fix proved straightforward (retrain with balanced data) but the deeper challenge involved recognizing how seemingly neutral algorithms perpetuate systemic discrimination.

Privacy concerns multiply when AI systems aggregate and analyze vast personal datasets. Leaders must balance AI's hunger for data with individuals' rights to privacy and autonomy. One financial institution developed a powerful fraud detection system that analyzed customers' location data, spending patterns, and social connections. While highly effective, it raised profound questions about surveillance and consent. The company ultimately implemented strict data minimization principles, using only essential information with explicit customer permission.

Transparency and explainability are at the heart of AI ethics. Black-box algorithms that make critical decisions without human-comprehensible reasoning undermine accountability. When an insurance company's AI denied coverage to seemingly qualified applicants, neither agents nor customers could understand why. This opacity not only frustrated stakeholders but potentially violated emerging regulations requiring algorithmic explainability. The company invested in developing interpretable models that could provide clear rationales for decisions.

Accountability frameworks must also be factored. When an AI system makes or influences a decision, responsibility cannot simply default to "the algorithm told me to." Leaders must establish clear chains of accountability, ensuring humans retain meaningful oversight of consequential decisions. One healthcare system requires physician sign-off on all AI-generated treatment recommendations, maintaining human responsibility while leveraging machine insights.

While automation may improve efficiency, leaders must consider their obligations to employees whose roles AI might eliminate. Progressive organizations view this as an opportunity for human development rather than mere cost reduction. When a major bank automated routine transaction processing, they invested saved resources into comprehensive reskilling programs, transforming tellers into financial advisors and relationship managers.

Homogeneous teams often share blind spots, missing potential biases or unintended consequences. One autonomous vehicle company learned this after their AI struggled to detect pedestrians with darker skin tones, a failing that more diverse development teams might have anticipated. They subsequently rebuilt their team to include ethicists, social scientists, and representatives from affected communities.

Many organizations establish AI ethics boards combining internal leaders with external experts in technology, ethics, law, and social impact. These bodies review high-stakes AI deployments, develop ethical guidelines, and provide ongoing oversight. Their independence from immediate business pressures enables more objective ethical assessment.

AI systems designed for one cultural context may violate norms elsewhere. A social media platform's content moderation AI, trained primarily on Western cultural standards, systematically misclassified legitimate political discourse in other regions as hate speech. Global organizations need ethical frameworks flexible enough to respect cultural differences while maintaining core principles.

When organizations deploy AI systems developed by vendors, they remain ethically responsible for outcomes. Due diligence must extend beyond technical performance to examine training data sources, built-in biases, and decision-making transparency. One retailer discovered their vendor's customer service chatbot had learned inappropriate responses from internet training data, requiring extensive review and retraining before deployment.

When competitive pressure pushes toward rapid deployment, ethical leaders insist on proper testing and bias assessment. When AI promises cost savings through aggressive optimization, they consider broader stakeholder impacts. This principled stance often proves profitable long-term, as organizations building trustworthy AI attract customers, talent, and partners who value integrity.

Organizations that view ethics as integral to AI development and not a compliance checkbox build systems that earn trust and deliver sustainable value. In an era where a single algorithm can affect millions of lives, ethical leadership becomes a strategic necessity. To ensure AI delivers on its promise without compromising values, organizations must track its real-world impact with equal rigor. That's where measurement becomes mission-critical.

MEASURING AND MANAGING AI IMPLEMENTATION SUCCESS

The biggest mistake in AI implementation? Measuring it like it's just another tech upgrade. The challenge isn't that AI lacks value, but that its benefits manifest in ways standard measurement frameworks fail to capture. Measuring AI implementation success with traditional ROI calculations is like trying to judge a chess grandmaster by how quickly they move the pieces; you're tracking the wrong variable. Organizations need measurement frameworks as sophisticated as the technology they're evaluating. Strategic leaders develop sophisticated approaches to tracking AI implementation that balance financial returns with transformative impact.

Before deploying predictive maintenance AI, a manufacturing plant must document current equipment failure rates, downtime costs, and maintenance efficiency. Without these benchmarks, claims of AI-driven improvements remain unsubstantiated opinions rather than measurable facts. One automotive manufacturer spent three months collecting granular baseline data before their AI rollout, enabling them to demonstrate 35 percent reduction in unplanned downtime with statistical confidence.

When a pharmaceutical company implemented AI for drug discovery, initial ROI calculations looked disappointing; the system hadn't yet produced a marketable drug. However, deeper analysis revealed the AI had eliminated 80 percent of molecular combinations that would have failed in testing, saving years of researcher time and millions in laboratory costs. Success measurement required shifting focus from outputs (drugs developed) to process improvements (research efficiency).

Financial metrics (cost savings, revenue growth) remain important but sit alongside operational measures (process efficiency, error reduction), customer indicators (satisfaction scores, engagement rates), and innovation metrics (new capabilities enabled, time-to-market improvements). This multifaceted approach prevents narrow optimization that achieves one goal while undermining others.

Some benefits appear immediately: an AI chatbot handling routine customer queries reduces call center load within weeks. Others emerge gradually: machine learning models improving through continuous training might take months to surpass human performance. Sophisticated measurement frameworks track both quick wins and long-term value creation, maintaining stakeholder support during the learning curve.

A healthcare system's AI diagnostic assistant showed impressive accuracy in tests but initially saw minimal physician adoption. Deeper investigation revealed the interface required too many clicks, disrupting clinical workflow. After redesign based on usage data, adoption soared to 90 percent, finally delivering projected benefits. Monitoring how humans interact with AI is as important as measuring the AI's technical performance.

An e-commerce platform's recommendation AI successfully increased average order values but also led to inventory imbalances as it oversold certain products. Only by tracking secondary effects (inventory turnover, stockout rates, supplier relationships) did leadership recognize and correct this optimization trap. Comprehensive measurement frameworks anticipate ripple effects beyond primary metrics.

Machine learning models drift over time as data patterns change, requiring adjusted performance benchmarks. A credit card company's fraud detection AI initially caught 95 percent of fraudulent transactions. As fraudsters adapted tactics, the rate dropped to 85 percent. Rather than viewing this as failure, the company tracked "adaptive performance" (how quickly the system learned new fraud patterns) revealing continuous improvement despite changing challenges.

Rather than tracking how many decisions AI makes autonomously, measure decision quality when humans and AI work together. A financial advisory firm found their AI investment's true value lay not in automated portfolio management but in how it enabled advisors to provide more personalized, timely recommendations to clients.

Industry averages for AI performance can mislead if organizations have different starting points, data quality, or implementation approaches. One retailer's inventory AI achieved "only" 15 percent accuracy improvement compared to the industry average of 25 percent, but starting from a more optimized baseline meant absolute performance exceeded peers.

When multiple improvements occur simultaneously (AI implementation alongside process redesign, training programs, and organizational restructuring) isolating AI's specific contribution becomes difficult. Statistical techniques like difference-in-differences analysis, where AI-enabled units are compared with control groups, provide more rigorous attribution, though it'll be short of perfect.

Regular updates that honestly present both successes and challenges demonstrate leadership commitment to transparency. One hospital system publishes monthly "AI Impact Reports" showing how clinical AI tools affect patient outcomes, staff efficiency, and cost structures. This openness has maintained board support through implementation challenges.

Initial metrics often focus on technical performance: accuracy rates, processing speeds, uptime. As implementations mature, measures shift toward business impact: customer lifetime value, market share gains, innovation velocity. The most advanced organizations track ecosystem effects - how AI enablement attracts talent, enables partnerships, or creates new business models.

Each metric provides feedback for refinement, each unexpected result offers learning opportunities. Organizations that view measurement as an integral part of AI development (not just post-implementation assessment) create virtuous cycles where insights drive enhancements, which generate better outcomes, producing richer insights.

The journey from AI investment to realized value rarely follows a straight line. Effective measurement systems provide navigation tools for this complex journey, helping leaders identify when to accelerate, when to pivot, and when to persist through temporary setbacks. In the AI era, the ability to measure and manage implementation success becomes a competitive advantage itself.

But measurement alone is a dashboard for a journey that never truly ends. In the fast-moving world of AI, today's breakthrough is tomorrow's baseline. To stay ahead, organizations must do more than track progress, they must continuously reinvent it.

SUSTAINING CONTINUOUS INNOVATION AND LEARNING

AI transformation requires perpetual evolution. The pace of AI advancement demands organizational clock speed that matches technological evolution. While Moore's Law described semiconductor improvements, AI capabilities grow even faster through algorithmic breakthroughs, expanding datasets, and novel applications.

Leading organizations revisit AI roadmaps quarterly, scanning for emerging technologies, evolving use cases, and shifting competitive dynamics. A retail bank that initially focused AI on fraud detection might discover opportunities in personalized financial advice or predictive customer service. Regular strategy updates ensure AI investments align with both current capabilities and future possibilities.

Each AI implementation generates insights about data quality requirements, integration challenges, change management approaches, and unexpected applications. Without systematic capture, these expensive lessons remain trapped in individual teams. Progressive organizations create AI centers of excellence that document patterns, share best practices, and accelerate subsequent implementations.

While optimizing existing AI systems delivers immediate value, breakthrough advances often emerge from experimental projects with uncertain outcomes. Following the 70-20-10 rule, organizations might allocate 70 percent of AI resources to enhancing current capabilities, 20 percent to adjacent opportunities, and 10 percent to moonshot experiments. This portfolio approach ensures steady improvement while maintaining optionality for disruption.

Technical skills evolve rapidly; yesterday's deep learning expert needs to understand today's transformer architectures and tomorrow's quantum machine learning. But sustainable advantage comes from developing innovation mindsets, not just technical capabilities. Organizations investing in curiosity, experimentation, and creative problem-solving build workforces that adapt to whatever technological shifts emerge.

Collaborations with universities provide access to cutting-edge research before commercialization. Startup partnerships inject entrepreneurial energy and novel approaches. Industry consortia enable pre-competitive knowledge sharing that advances entire sectors. One pharmaceutical company's AI research partnership with a quantum computing startup positions them for next-generation drug discovery capabilities that won't mature for years.

When an insurance company's AI underwriting system produced unexpected results, they hosted a "failure party" where the team shared learnings that informed three successful subsequent projects. Organizations that stigmatize failure discourage the experimentation necessary for breakthrough innovation. Those that mine failures for insights accelerate learning curves.

Tracking the number of AI experiments launched, the speed of iteration cycles, the percentage of employees contributing ideas, and the time from concept to pilot provides leading indicators of innovation health. One technology company measures "innovation metabolism", or how quickly ideas flow from inception through testing to implementation or abandonment.

Customer involvement in innovation creates market-aligned evolution rather than technology for technology's sake. When a healthcare AI company invited nurses to their development sprints, the resulting features addressed workflow realities that engineers never imagined. The head of product development noted wryly, "Turns out, actual users have insights our PhDs missed. Who knew?"

Regular hackathons where business teams partner with data scientists to prototype solutions. Innovation showcases where employees present creative AI applications, no matter how quirky. Innovation doesn't always look like a TED Talk. Sometimes it looks like an accounting firm's 'AI Weird Science Fair.' One team built a machine learning model to predict coffee machine breakdowns (because nothing impacts accountant productivity quite like a caffeine outage). This is the sweet spot of AI adoption: solving problems that actually annoy people. If you let your team use AI to solve their petty grievances, you'll be amazed at how quickly they master the technology.

Organizations limiting themselves to local innovation miss transformative opportunities. Talent pools, research breakthroughs, and novel applications emerge globally. One European manufacturer's partnership with an AI lab in Singapore unlocked manufacturing optimizations their domestic teams hadn't considered. The plant manager admitted, "Sometimes the best ideas come from people who don't know what's 'impossible' in your industry."

The compound effect of continuous innovation becomes visible over time. Accumulated knowledge, refined processes, and innovation culture create widening competitive moats. A logistics company that began with simple route optimization now operates an AI ecosystem orchestrating everything from demand prediction to autonomous vehicle deployment.

Sustained innovation requires emotional resilience alongside technical capability. The AI journey includes setbacks, surprises, and occasional existential crises about human relevance. Leaders who acknowledge these challenges while maintaining optimistic persistence create organizations that thrive amid uncertainty. As one battle-tested CTO advised, "Expect your first three AI projects to teach you primarily what not to do. The learning is worth more than the immediate results."

The paradox of AI innovation lies in its simultaneous acceleration and democratization. Capabilities that required millions of dollars and PhD teams five years ago now run on cloud platforms accessible to any developer. This democratization means competitive advantage increasingly flows not to those with the most resources, but to those who learn and adapt fastest. Organizations building continuous innovation capabilities position themselves to capture value from whatever emerges next in AI's evolution.

SUMMARY

The AI transformation challenge extends beyond selecting models and deploying code. Leaders face more profound questions: How might we reimagine organizations where humans and machines amplify each other's strengths? How can we preserve our values while embracing algorithmic decision-making? And most critically, how do we navigate this transition while keeping people at the center?

Successful AI-era leadership begins with strategic clarity - connecting technological capability to organizational purpose with precision. It requires a recalibrated skillset where technical literacy complements emotional intelligence, enabling leaders to guide anxious teams toward collaborative partnership with intelligent machines. Transparent communication transforms resistance into engagement, while cultural adaptability turns disruption into evolution. Ethical boundaries prevent algorithms from amplifying biases or compromising values. Sophisticated measurement looks beyond simplistic ROI toward multidimensional value creation.

Perhaps most importantly, sustainable transformation demands continuous learning: recognizing that AI implementation isn't a destination but a perpetual journey of advancement. Organizations embedding innovation into their cultural DNA will help build the road.

The more powerful our technology becomes, the more essential our humanity grows. Algorithms may optimize efficiency, but purpose, creativity, and ethical judgment remain uniquely human domains. Leaders who recognize this, and who view AI as amplifying human potential rather than replacing it, create organizations that thrive amid automation while preserving what makes them distinctly human.

Chapter takeaways:

- Successful AI transformation demands paradoxical leadership capabilities: bold vision with patient execution, technical sophistication with human empathy, strategic clarity with adaptive flexibility.
- The human element threads through every aspect of successful AI transformation; technology may be the catalyst, but people remain the constant.
- Leaders who master orchestrating human and artificial intelligence will define the future of organizational success.

As we turn toward Chapter 4, we shift from organizational architecture to the fundamental building blocks of human-AI partnership. How might we redesign individual roles, team structures, and workflows to create truly symbiotic relationships between people and intelligent systems? How do we reimagine work itself when machines handle routine tasks and humans focus on judgment, creativity, and interpersonal connection? These questions await as our journey continues: from transforming organizations to reinventing work itself in an age of artificial intelligence.

4

EMPOWERING TEAMS AND INDIVIDUALS IN AN AI-ENHANCED WORKPLACE

AI transforms how we work at every level. How will you empower teams and individuals as AI assumes greater workplace dominance?

This chapter reveals how AI can amplify productivity, creativity, and decision-making when implemented thoughtfully. Healthcare, finance, and IT examples demonstrate how AI assists professionals in delivering superior outcomes. We explore career development and employee growth in the AI age through reskilling programs, new career pathways, and continuous learning initiatives that help employees adapt and seize AI-created opportunities.

We examine managing employee engagement while tackling job security concerns, workload stress, and "technostress." You'll discover approaches to mitigate fear and resistance, communicate transparently about AI changes, and leverage AI itself to improve work-life balance and satisfaction.

Organizations must remain agile through continuous feedback and improvement in human-AI interactions. Establishing feedback loops between humans and AI drives ongoing learning on both sides, spurring better performance and innovation. You'll understand how to create environments where humans and AI continuously help each other improve. First, we'll explore how organizations are reconfiguring work to foster deeper human-machine collaboration.

HUMANS AS SENSEMAKERS IN AI LOOPS

The phrase "human-in-the-loop" has become shorthand for AI safety and accountability, but in high-functioning AI-enhanced organizations, the human role extends well beyond checking for hallucinations or approving outputs. AI excels at first-order cognition: finding patterns, summarizing data, generating plausible responses. But true organizational value is created through second-order thinking: asking better questions, abstracting implications, and connecting insights across time and domain. AI can process a billion data points in a second, but it still struggles to understand why sending an invoice at 4:59 PM on a Friday is a hostile act. This is where the human comes in. We are the 'Sensemakers.' Our job is to look at the AI's technically correct output and say, 'Yes, but if we send that to the client, they will fire us.' AI provides the text; humans provide the subtext. AI provides the data; humans provide the apology.

In this context, employees are interpreters and recombiners. They decide which AI outputs matter, where judgment must override automation, and how emerging insights should shape future direction. They draw on lived experience and organizational context to make sensemaking a core capability. And it reframes workforce development away from "learning how to use the tool" toward developing judgment and narrative-building. Teams must learn not only how to prompt AI, but how to challenge its framing, compare its outputs, and refine its usefulness through iterative interpretation. When this capacity is embedded in team culture, AI becomes a stimulus for better thinking. And organizations evolve through continuous reframing.

The World Economic Forum's Future of Jobs Report 2025 projects that by 2030 around 92 million jobs could be displaced due to automation, AI and other labor-market trends, while approximately 170 million new roles could emerge globally over the same period — resulting in a net gain of about 78 million jobs. As generative AI displaces large segments of routine knowledge work, a new organizational elite is emerging: not by title or tenure, but by their ability to convert ambiguity into insight and into action. These knowledge arbitrageurs aren't valuable because they know more. They're valuable because they make more sense of what's already known. They interpret across contexts. They frame relevance. They ask the right questions before anyone else knows answers are needed. These roles often resist traditional job descriptions. They include:

Product leads who turn tech changes into business choices: When the model slows, costs spike, or security rules tighten, they explain what that means for sales, profit, and the roadmap - and make clear calls to ship, pause, or stop before the tech quietly sets the strategy.

Field operators who spot breakdowns early: They notice repeating workarounds, hesitations, or customer questions, bring a few concrete examples, run a quick safe test, and confirm by day's end whether a small issue is really a system problem.

Strategists who connect weak signals into foresight: They link small clues (pricing shifts, hiring spikes, draft rules) into a few likely futures, name the signs to watch, and propose no-regret moves the business can fund now.

Change agents who turn local wins into reusable playbooks: They capture what actually worked (prompts, checklists, "works/doesn't" notes), test it in a couple of new teams, trim the edge cases, and publish a simple kit others can adopt in days, not months.

Without them, AI output becomes noise, dashboards become wallpaper, and decisions become fragile.

Leaders must help organizations do four things well:

- Spot arbitrage talent early: Look for people at the edges of roles or disciplines who connect dots, ask sharp questions, and pilot small tests. Give them a name, a sponsor, and a path.
- Give them flexible roles and room to roam: Let them move across teams, pull data, interview users, and recombine ideas so insights travel faster than the org chart.
- Make their thinking visible: Ask them to show their method: assumptions, sources, quick tests, and how they reached a call. Capture it in short, reusable notes or templates so others can learn and copy the moves.
- Win by keeping and multiplying your best sense makers: Retain them, reward them, and have them teach others. In an automated world, the edge goes to firms that grow more people who can make sense and help the rest of the system think that way.

In the age of automation, the battle for competitive advantage will come down to who keeps their most adaptive sense makers and how they're empowered to teach the rest of the system to think in kind.

Focus on imagination and re-imagination of work, fundamentally redesigning workflows so AI and people collaborate at what each does best. Many executives approach AI integration like medieval alchemists, convinced they can transform base metals into gold with the right incantation. The reality is less magical and more methodical: successful AI transformation isn't about finding a philosopher's stone, but about carefully redesigning the entire workshop.

Consider a bank's finance department transformation. Instead of clerks manually compiling reports, employees now act as AI-enabled analysts, overseeing AI-generated routine reports, validating results, and focusing on advising business units with AI-surfaced insights. The role shifts from data gathering to supervision and strategic analysis.

In HR departments, staff answering routine employee queries now work alongside AI chatbots. The chatbots handle bulk questions while HR professionals oversee the bot, update its knowledge base, and tackle complex, sensitive human issues. This transforms repetitive support into strategic positions requiring decision-making and AI management skills.

A global pharmaceutical company rolled out a generative AI tool to help its field sales reps quickly summarize clinical trial data for physician conversations. At first, the rollout focused on training and adoption to teach reps how to generate summaries using the AI assistant. Instead of simply copying and pasting AI outputs into their scripts, several reps began customizing the AI summaries based on physician type, regional concerns, and treatment priorities. One rep built a library of prompts tailored to different provider personas: oncologists who wanted statistical rigor, primary care doctors who preferred plain-language summaries, and pharmacists who needed risk profiles. These customized prompts started circulating informally among teams. Reps began comparing which prompts generated the most effective conversations and tweaking them based on real-world feedback. The change team noticed this grassroots behavior and shifted strategy. Instead of enforcing one standardized workflow, they launched a knowledge-sharing sprint: reps shared high-performing prompt patterns, created a "what worked, where" heat map, and published a lightweight internal library of AI-generated talking tracks with physician feedback loops. In the end, the AI made reps smarter and more adaptable. The company saw measurable improvements in call quality and follow-up effectiveness. But more importantly, it sparked a new mindset: reps were co-creating insight in real time and turning everyday interactions into a scalable knowledge asset.

Collaborative approaches see better results: According to published reports, Accenture research found that companies fixated on using AI purely for cost reduction often stall, while those redesigning work to amplify human strengths achieve greater innovation and growth. Ask not "What jobs can we automate?" but "How can we redefine jobs so our AI-supported people accomplish previously impossible things?"

Middle managers and team leads must collaborate with frontline employees to identify tasks best suited for AI versus those requiring human touch. Involving employees in redesign reduces fear; people see AI as a tool in their toolkit, not a job competitor. To fully realize the benefits of collaborative AI, organizations must go beyond redesigning roles; they must also equip their workforce with the skills to thrive in these new roles. While leadership and employee collaboration are key to redefining work, success hinges on democratizing AI knowledge across the organization.

DEMOCRATIZING AI KNOWLEDGE ACROSS YOUR ORGANIZATION

Redesigning roles only succeeds when employees possess the knowledge and skills for new human-AI collaborative positions.

A 2023 HR leaders survey found more than half of organizations now consider AI literacy a must-have skill for all employees. Workers who effectively use AI tools will outperform those who cannot. A Harvard Business Review analysis highlighted that "humans with AI will replace humans without AI," emphasizing the career risk of failing to adapt. Research shows employees most vulnerable to job displacement are those who don't engage with AI at all, while those learning to leverage AI become more valuable.

AI literacy doesn't mean converting every employee into a data scientist. It involves providing practical understanding of how AI works, where it applies in one's job, and how to interact with AI tools effectively and ethically.

Employees should understand what AI and machine learning are in simple terms, the tasks AI excels at (pattern recognition, predictions) and its limitations (hallucinations, potential biases). Banks might run workshops explaining how AI loan review systems analyze historical data to flag risky applications but lack human insight for special circumstances (where loan officer judgment remains crucial).

Organizations deploying AI-powered software need day-to-day usage training. This includes crafting effective queries for optimal results (prompt engineering), interpreting AI outputs, and validating accuracy. Marketing teams using AI copywriting tools need training beyond clicking "generate"; they must craft effective prompts and review AI-suggested copy for factual correctness and brand alignment. Training needs to emphasize AI as assistant, not infallible oracle.

Foster ongoing upskilling through regular micro-learning sessions, e-learning modules, or peer knowledge-sharing groups. One innovative approach: creating small learning "pods" where four-employee groups experiment hands-on with AI tools on work and non-work tasks, sharing experiences and collectively learning effective usage. Managers should train alongside teams for better support and guidance.

AI literacy training must include ethical and responsible use. Employees need understanding of data privacy, algorithmic bias, and company AI usage policies. Healthcare providers rolling out AI diagnostic aids must train staff not just on tool usage but on proper patient data handling and knowing when to override AI suggestions for patient care.

Many companies begin with awareness seminars and curated online courses for all staff. Pairing digitally native junior employees with senior staff as tech buddies spreads AI skills hands-on. Make AI training both a perk and talent attraction selling point, emphasizing cutting-edge skills development opportunity. This widespread literacy sets the stage for deeper human-AI collaboration. People who understand technology are more likely to trust it and find innovative partnership ways.

BUILDING TRUST IN HUMAN-MACHINE TEAMS

AI-enhanced workplaces reach full potential only when built on inclusion, trust, and openness.

Leaders must proactively explain what AI is, what it will do, and what it won't do. When rolling out AI analytics platforms, communicate: "This AI helps analyze campaign data faster, allowing marketers more time for creative strategy; it assists rather than replaces you." Framing AI as augmenting abilities reduces automation anxiety. Formalize this into "Employee AI Manifestos" or digital empowerment charters, reassuring employees with principles like: no AI implementation without clear employee or customer benefit, and workers receive reskilling opportunities for new technology adoption.

Cross-functional teams piloting AI projects should mix AI specialists with end-users from various departments. A hospital implementing AI scheduling formed committees including doctors, nurses, administrative staff, and IT personnel. This committee provided design input ensuring real-world workflow accommodation, tested early versions, and championed benefits to peers. Such approaches surface potential issues early while giving employees ownership.

Nothing causes organizational anxiety quite like an ambiguous email about 'exciting AI efficiencies coming soon.' It ranks just above 'We need to talk' and slightly below 'Surprise audit tomorrow' in the pantheon of corporate terror. Trust comes from understanding and reliability. Organizations should make AI decisions transparent and explainable wherever possible. If employees rely on AI recommendations, they need rational understanding. Technical teams must ensure AI tools remain reliable and unbiased through regular accuracy and fairness audits plus ethical guidelines. Introduce AI ethics committees or business unit "AI ambassadors": employees trained to monitor departmental AI usage and address ethical or fairness concerns. Such governance helps all team members feel safer and included.

Because AI operates continuously doesn't mean expecting humans to become machines. Leaders must set realistic workload and pace expectations after AI introduction. If AI creates efficiency gains, invest that time in creativity or quality improvement, not simply loading more work on people. Younger employees might adapt quickly to AI tools while veterans feel less comfortable. Inclusive approaches might pair them in bidirectional mentoring relationships: younger employees help seniors with AI tool usage while seniors share domain experience and critical thinking approaches.

If performance evaluations incorporate AI analytics, validate that metrics truly correlate with performance without built-in biases. Adjust AI systems to account for different work styles or provide alternative contribution demonstration methods.

Focusing on openness, participation, and trust creates environments where employees feel that they are on the same team as AI, working toward common goals rather than competing with robots. An inclusive culture encourages employees voicing AI concerns and suggesting improvements, strengthening implementation while ensuring AI benefits are widely shared.

AMPLIFYING HUMAN POTENTIAL THROUGH MACHINE INTELLIGENCE

AI augmentation represents the heart of an AI-enhanced workplace: using tools to amplify and extend human capabilities.

Published research from organizations like the Boston Consulting Group and Harvard Business School confirms that employees using generative AI can achieve significant productivity and quality gains across real-world business scenarios. A 2023 study found that consultants using AI completed 12.2 percent more tasks, 25.1 percent faster, and with 40 percent higher quality than their peers. The impact varies by role: software developers using AI coding assistants generally experience substantial increases in output, while customer support agents assisted by chatbots handle 14 percent more hourly inquiries, especially newer workers.

Legal teams reviewing contracts with AI support scanning for risk clauses allow lawyers processing many more weekly contracts. Accounting departments using AI auto-classifying expenses and detecting anomalies free accountants for complex financial analysis. AI-suggested responses in customer service made agents faster while ensuring consistent best-practice language, improving service quality. Medical AI diagnostic assistants identify early disease signs human clinicians might miss, effectively providing second eyes raising overall diagnosis accuracy.

Junior consultants using AI research assistants quickly gather and summarize client market data. AI handling grunt work allows junior analysts producing insights rivaling senior consultant deliverables. Software development showed biggest productivity boosts in cognitively demanding tasks, with relative improvement greatest for those with the most to gain. Novices benefited enormously following AI code suggestions, learning better coding patterns.

Architecture firms saw junior architects spending days iterating design variations. Now, generative AI produces and tests variations in minutes based on architect-provided parameters. Human architects focus on curating promising designs and adding artistic touches AI cannot provide.

Marketing creative teams use AI generating draft visuals or taglines as inspiration for team improvement, resulting in more innovative campaigns. AI handles high-volume, high-speed generation while humans apply judgment, aesthetic sense, and contextual understanding.

A portfolio manager traditionally spent hours sifting through financial news and performance data. Today, AI scans global data continuously, flagging relevant developments. Managers receive AI-curated briefings highlighting portfolio concerns or opportunities, focusing expertise on verifying signals and devising strategic responses. AI extends awareness and analytical reach, monitoring more information than humanly possible alone for more informed decisions.

In a modern office buzzing with both people and algorithms, the vision of human-computer partnership comes alive. Rather than AI taking over, it's woven thoughtfully into daily work, amplifying human strengths. Think of an advisor at a financial services firm: each morning the advisor's AI assistant sifts through research and even drafts a summary of yesterday's client meeting. By the time the advisor sips a first coffee, routine data-crunching is done – freeing them to focus on nuanced strategy and client relationships. A leading financial services firm embraced this approach in 2024, rolling out an AI "knowledge assistant" and a meeting notes bot that don't replace the advisors, but supercharge their preparation. Managers there report that nearly all their advisors use these tools, because they save time without eroding the human touch. It's a scene straight out of J.C.R. Licklider's "Man-Computer Symbiosis" vision from 1960: humans and machines working in tandem, each doing what they do best. The AI handles the tedious chores and provides real-time insights, while the humans apply judgment, creativity, and empathy. Crucially, leadership guided this integration with care – training staff, refining workflows, and making the AI interface so user-friendly that it feels like a natural extension of the team. The result is a balanced partnership where employees stay engaged and never feel overshadowed by automation. In this narrative of change, AI isn't a threat lurking in the server racks; it's a colleague that shoulder-to-shoulder helps solve problems, accelerating projects without stealing the thrill of human decision-making. Such stories show how, with the right change management, Licklider's symbiotic dream is becoming a practical reality in workplaces today.

FUTURE-PROOFING CAREERS IN THE MACHINE AGE

Empowering individual employees extends beyond current job improvement - it ensures organizational futures.

In the face of rapid technological change, recent global analyses highlight an urgent upskilling imperative for today's workforce. The World Economic Forum's Future of Jobs research estimates that about 60 percent of workers will require reskilling or upskilling by 2027, with roughly 44 percent of an individual's core skills needing an update to keep pace with evolving job demands. Organizations worldwide are responding by making talent development a strategic priority rather than an afterthought. Indeed, 85 percent of employers now plan to prioritize reskilling their employees, reflecting a decisive shift in corporate training budgets from peripheral spending to essential investment. Many leading companies have launched internal "reskilling revolutions": large-scale programs to equip their people with new skills for future roles, underscoring a global consensus that upskilling current employees is key to remaining competitive in the age of AI and automation.

Upskilling trains new skills within current domains, while reskilling prepares employees for entirely different roles. Administrative assistants whose routine tasks become AI-handled need upskilling in advanced data analysis or project coordination for higher-value responsibilities. Warehouse workers might reskill into robotics maintenance technician roles as companies deploy automated systems. This can be accomplished by partnering with educational institutions or offering in-house certification programs.

When assessing AI adoption skill gaps, what new roles emerge? What skills do those require? Which current roles decline? Clear skill roadmaps enable targeted individual development. By creating personalized learning journeys; employees receive "skills profiles" and development plans outlining needed training as roles evolve. IT departments implementing AI might welcome customer service employee rotation for project learning, bringing knowledge back to home departments. Cross-functional projects can spark new career direction interest. AT&T famously encouraged employees charting new career paths when anticipating massive skill shifts, providing sizable continuous learning budgets. Amazon publicly committed to a $1 billion+ workforce upskilling investment, offering hundreds of thousands of employees tech role training through 2025.

In today's AI-augmented workplace, career development conversations are evolving. Managers explore how AI is reshaping roles and spotlight the skills that will matter most tomorrow. They help employees map out learning journeys with purpose. Senior leaders lead by example, showing that curiosity and adaptability outweigh any current gaps in AI fluency. For junior lawyers: instead of grinding through hours of document review, they're stepping into client meetings and strategy sessions earlier in their careers. The time saved by AI is reinvested in growth. That freed-up capacity becomes a launchpad for deeper learning, faster confidence-building, and accelerated paths to leadership.

Career paths are branching in new directions. Organizations spotlight emerging roles and invite employees to explore possibilities. Internal job fairs showcase AI-involved futures, helping people envision what's next. Supportive cultures reduce the stigma of role change. Employees gain confidence knowing AI-affected positions come with guidance, not abandonment. Trust grows when companies commit to retraining before any automation-driven layoffs.

Leadership development is shifting toward human strengths. Creativity, emotional intelligence, and cross-cultural collaboration take center stage. Employees who blend digital fluency with empathy and ethics rise faster. This new mix creates "fusion skills": the ability to pair AI-powered workflows with sound judgment. These hybrid capabilities define tomorrow's leaders.

PRESERVING HUMAN VITALITY WHILE MACHINES JOIN THE TEAM

When deployed thoughtfully, AI removes soul-crushing drudgery, introduces flexibility, and improves work-life balance. Yet the same technology, when mismanaged, triggers anxiety, resistance, and alienation that can undermine your transformation efforts entirely.

Address Concerns Head-On Before They Metastasize

When employees hear that AI is coming to their department, their first instinct is rarely excitement about reduced busywork. More often, it's fear - fear of redundancy, irrelevance, or being replaced. This reaction isn't irrational. It's human. Leaders who dismiss these concerns as mere "resistance to change" miss a vital opportunity to build trust and shape the narrative.

Proactive communication is essential. Structured forums (town halls, small-group discussions, anonymous question portals) create space for employees to ask the questions they're actually thinking: Will my job exist in a year? Are we just training the system that will replace us? How will my day-to-day work change? They're calls for clarity. When Microsoft introduced its AI Copilot, it demonstrated how the technology would support specific tasks and explicitly affirmed that human expertise remained central. That transparency turned skepticism into cautious optimism.

In organizational life, silence is rarely neutral. The less clearly leaders communicate, the more creatively employees fill in the blanks. Rumors thrive in ambiguity. Even the most analytical teams will construct elaborate theories when left in the dark. Nature abhors a vacuum, but workplace culture actively resents it. And if leaders don't shape the story, someone else will.

Redesigning Work to Preserve Meaning

Research consistently shows that AI's impact on employee engagement depends less on whether it changes the nature of work, and more on how it changes the meaning behind it. When AI removes tasks that provide pride, purpose, or fulfillment without enriching the role elsewhere, disengagement is not just likely, it's inevitable.

Consider the relationship manager whose client interactions form the core of their professional identity. If those interactions are suddenly delegated to an AI system, the manager may remain employed, but their sense of purpose disappears. Contrast that with a scenario where AI handles routine documentation, freeing the same manager to deepen client relationships and focus on complex problem-solving. The technology is the same. The engagement outcome is dramatically different.

The question isn't whether AI will take over tasks currently performed by humans. The strategic question is: which tasks should we delegate, and how do we redesign roles to make them more meaningful in the process? Leaders who treat AI as a tool for role enhancement and not just task replacement are far more likely to sustain engagement and unlock long-term value.

Allstate's Insurance Claims Transformation

According to public reports, when Allstate introduced AI-powered claims processing, the goal was clear: speed up resolution, reduce manual effort, and improve customer experience. The technology (capable of analyzing damage photos, estimating repairs, and flagging potential fraud) dramatically shortened turnaround times, often processing simple claims in hours instead of days.

For claims adjusters, the shift was significant. Many had built careers around evaluating damage and detecting fraud - skills now partially automated. Allstate's leadership recognized the potential for anxiety and proactively invested in retraining programs, allocating tens of millions toward helping employees adapt to new roles. Human oversight remained central: adjusters continued to review complex claims and had final authority over AI-generated decisions. In practice, this meant AI handled routine tasks, while adjusters focused on nuanced judgment calls and customer interactions.

Rather than framing the initiative as automation-led, Allstate emphasized partnership. AI was positioned as a "digital colleague," supporting human expertise. Claims agents reviewed every AI-generated communication before it reached customers, ensuring empathy and accuracy. Allstate reported improved customer satisfaction and faster service when humans and AI worked together.

The lesson: meaningful integration of AI depends not just on technical capability, but on preserving employee purpose and ensuring human judgment remains at the core.

Manage the Pressure Paradox

AI's efficiency creates an unexpected challenge: when systems perform dramatically faster, human capacity is often strained to keep up. As organizations adopt AI tools to streamline workflows, expectations often rise even faster than the technology can deliver. Without intentional boundaries, the productivity gains from automation are quickly offset by expanded workloads, creating a paradox where technological progress intensifies workplace stress.

Forward-thinking companies recognize this risk and implement guardrails around AI-enabled work. For example, if AI reduces meeting coordination time by 70 percent, that reclaimed time should be reserved for strategic thinking or creative exploration, not simply filled with more tasks. This requires discipline. The instinct to raise performance targets in response to efficiency gains is strong, but unchecked, it erodes employee well-being and long-term productivity.

Another emerging challenge is "technostress": the anxiety and frustration that comes with adapting to new systems. Navigating this learning curve demands psychological safety. Employees must feel free to acknowledge their struggles without fear of judgment. Leaders play a critical role here: by openly sharing their own experiences with technology adoption, they normalize the discomfort and foster a culture of resilience and empathy.

Maintain the Human Touch in System Design

Few things damage engagement faster than feeling evaluated by algorithms you don't understand. When AI enters performance management processes, maintaining human judgment becomes critical. Pure algorithmic evaluation feels dehumanizing, especially when employees suspect the system misses important nuances of their work.

Progressive organizations use AI to provide performance data points, but keep managers firmly in control of holistic evaluation discussions. This reinforces that employees are valued as individuals with context and complexity, not just data points in a system.

Similarly, human-centered design principles should guide all AI implementations. Every interface, workflow, and process should reflect the question: "How would I feel using this system?" Organizations that test AI tools with diverse employee groups before full deployment catch engagement-damaging design flaws early.

Build Feedback Cycles That Matter

Casual check-ins like "How's the new system working for you?" rarely surface meaningful insights. Employees often hesitate to share real concerns in informal settings, especially when the stakes involve performance or job security. Instead, leaders must build structured feedback mechanisms (such as AI experience surveys, facilitated retrospectives, and anonymous input channels) that invite honest reflection and capture how technology is truly affecting daily work.

When an employee reports that "the AI scheduling system keeps suggesting unrealistic deadlines," that's a signal. It's product intelligence. Organizations that treat this kind of feedback as essential input for system refinement improve both their technology and their culture. They demonstrate that employee experience is not a side effect of innovation, it's a core metric.

But feedback alone isn't enough. Closing the loop is what builds trust. When employees see their input lead to tangible system changes, they shift from passive users to active co-creators. This partnership mindset turns resistance into collaboration and transforms implementation into shared ownership.

Even the most advanced AI deployment will falter if it ignores the human experience. Leaders must address concerns directly, preserve meaningful work, manage pressure thoughtfully, and maintain human touchpoints in digital systems. Most importantly, they must build feedback mechanisms that are not just functional, but consequential. That's how organizations create the conditions for people and machines to thrive together.

ENGINEERING A SYMBIOTIC FUTURE

Successful evolution requires a cultural architecture where continuous improvement becomes reflexive for both human and artificial intelligence.

Most organizations treat AI implementation like marriage in a Jane Austen novel - once the ceremony is complete, the story ends. The reality is closer to modern relationships: the ceremony merely begins ongoing negotiation, adjustment, and occasionally, couples therapy. Organizations that thrive aren't those implementing AI perfectly the first time; they're the ones building robust feedback systems for when things inevitably go sideways.

This evolutionary mindset represents a fundamental paradigm shift. Rather than measuring success by initial deployment metrics, forward-thinking leaders evaluate their AI initiatives by how effectively they learn and adapt over time. The maturity question shifts from "How sophisticated is your AI?" to "How rapidly does your AI-human system improve?"

The most elegant feedback architectures embed learning mechanisms directly into the AI tools. Consider a simple decision-support AI that recommends pricing strategies. Rather than assuming its recommendations are inherently valuable, sophisticated versions allow users to rate suggestions immediately: Was this helpful? Did it miss important context? Would you implement this recommendation? This real-time evaluation data becomes training input for the next iteration. The system doesn't just accumulate data; it develops judgment through human guidance. The beauty of this approach is its transparency - employees aren't mysterious subjects being observed by the AI but active collaborators in its development. More profoundly, these feedback loops shift the power dynamic. Employees become teachers rather than users, developing psychological ownership in the AI's success. When a sales representative sees how her feedback directly improved next week's AI recommendations, she transitions from technology consumer to co-creator.

When a financial services firm implemented an AI platform to analyze legal documents, the initial results were technically impressive but operationally mixed. The system could process thousands of documents in minutes rather than the many many hours previously required by lawyers and loan officers.

However, after the initial deployment euphoria, frustration emerged among legal teams. The AI excelled at standard agreements but struggled with nuanced contracts, creating extra work for attorneys who had to double-check everything. The system seemed promising but incomplete, and improvement was painfully slow.

Rather than retreating to traditional methods or forcing adoption, the firm reimagined their approach. They:

Created a "legal issue flagging" interface directly within the document review system where attorneys could highlight contracts where AI interpretations were questionable, capturing not just the error but the contextual reasoning.

Established weekly "learning rounds" where data scientists and legal experts reviewed flagged cases together, combining technical and domain expertise to refine the algorithm.

Built a performance visualization dashboard showing how specific legal teams' feedback was improving the AI's accuracy in their domains of expertise, creating healthy competition among teams to "teach" the system better.

Instituted a "contract complexity bonus" where attorneys received recognition for helping the AI navigate especially difficult legal scenarios.

Within eight months, AI accuracy on complex agreements improved from roughly 60 percent to more than 90 percent. More significantly, lawyer satisfaction with the system rose significantly. Many attorneys who had initially resisted the system became its strongest advocates. The feedback process had transformed lawyers from reluctant users into invested teachers. The evolution continues today with legal teams and AI engaged in continuous mutual improvement. By institutionalizing feedback and celebrating human expertise, the firm created a symbiotic system that exceeds the capabilities of either humans or AI operating independently.

Build Cross-Functional Evolution Teams

Siloed improvement efforts inevitably create siloed results. When AI systems run, the technical teams that build them must maintain constant contact with the business teams using them. Feedback loops are human connections: people talking regularly about what works and what doesn't.

A monthly meeting between AI developers and end-users represents the minimum viable communication structure. These conversations surface qualitative insights impossible to capture in metrics: "The dashboard is too confusing," or "We'd like the AI to also do X if possible." Technical teams provide intelligence about new capabilities, while users provide intelligence about emerging needs.

The organizations making the most dramatic progress treat their AI systems as products with cross-functional ownership. Health insurers like Kaiser Permanente have pioneered this approach, creating "AI product managers" who bridge technical and operational teams, ensuring steady feedback flow and shared accountability for outcomes.

Monitor Both System and Human Performance

Most AI implementers track obvious technical metrics: processing speed, error rates, cost savings. More sophisticated organizations simultaneously monitor human experience metrics: employee satisfaction with AI tools, perception of workload impact, and sense of control over AI-augmented work.

This balanced dashboard acknowledges that technical performance without human engagement creates the equivalent of a high-performance sports car with no driver - impressive in theory, useless in practice. When turnaround time improves but employee satisfaction plummets, that signals the AI may be pushing too hard or feels cumbersome to use.

Leaders should periodically ask: How do employees in AI-augmented roles feel about their jobs? Are they more engaged or less engaged than before? Which aspects of the AI partnership feel empowering versus frustrating? These insights predict long-term success better than any processing metric.

Normalize Continuous Iteration

Perfectionism kills evolution. Organizations must normalize iteration as the expected pattern, not a sign of initial failure. When an AI scheduling system causes unforeseen manufacturing conflicts because it doesn't understand tacit shop-floor constraints, leaders should frame this as valuable discovery rather than implementation failure.

This mindset shift requires explicit communication: "Our feedback directly drives improvements." When employees see problematic AI suggestions corrected after they raise concerns, it validates the process and encourages more proactive participation in system refinement.

Some organizations take this further by explicitly rewarding employees who identify improvement opportunities. One pharmaceutical company awards quarterly "AI evolution prizes" to teams that discover non-obvious ways to enhance their drug discovery algorithms, creating positive reinforcement for critical feedback.

Encourage Experimental Intelligence

Beyond organized feedback, organizations should encourage exploratory AI usage. Employees often discover novel tool applications that developers never envisioned. These unexpected use cases frequently drive the most significant long-term value.

Financial service firms that initially deployed natural language processing for routine customer inquiries discovered that advisors were using the same technology to prepare for client meetings: creating personalized research summaries that previously would have taken hours. Rather than discouraging this unplanned usage, they expanded API access and created template prompts specifically for this purpose.

By creating safe spaces for experimentation and celebrating creative applications, organizations tap into collective intelligence far exceeding what any implementation team could design. The most valuable AI adaptations often emerge from the frontlines, not the development lab.

The future belongs to organizations that build not just intelligent machines but intelligent relationships between humans and machines. By focusing on continuous evolution rather than perfect implementation, companies create adaptive systems that improve organically over time. Like any successful partnership, the human-AI relationship thrives not because of flawless beginnings but because of commitment to mutual growth.

SUMMARY

The successful AI-enhanced organization operates neither as disconnected parts nor perfect harmony, but as a deliberate frontier where human ingenuity and machine capabilities constantly redraw their boundaries. Leaders who view AI implementation as a destination rather than an ongoing expedition inevitably create static solutions for dynamic challenges.

This frontier approach requires a cultural architecture where exploration and adaptation become reflexive. Organizations must dismantle the "set and forget" fallacy by embedding feedback mechanisms directly into AI tools, creating cross-functional discovery teams, and balancing system performance with human experience metrics.

This philosophy is exemplified when organizations transform professionals from hesitant users into engaged innovators by establishing intentional feedback loops and recognizing their contributions. Such an approach leads to significant improvements in both AI system accuracy and user satisfaction. These experiences confirm that the most impactful AI implementations are not just technically advanced —they are also socially intelligent.

The most adaptive organizations encourage experimental intelligence, normalizing iteration as expected rather than failure. They recognize that frontline employees often discover the most valuable AI applications, and create environments where this creativity flourishes. This continuous evolution mindset creates resilient systems that improve organically in response to changing conditions, the hallmark of truly intelligent enterprise.

Chapter takeaways:

Redefine work around complementary strengths – Design roles where AI handles data-intensive tasks while humans focus on judgment, creativity, and relationships

Democratize AI literacy across all levels – Ensure everyone understands AI basics, practical applications, and ethical implications in their domain

Build trust through transparency and inclusion – Create environments where employees participate in AI implementation decisions and understand how systems work

Amplify human capabilities rather than replace them – Deploy AI as force-multipliers that expand what employees can accomplish, not as substitutes

Future-proof careers with continuous development – Establish clear reskilling pathways that help employees evolve alongside technology

Preserve meaning and purpose amid automation – Carefully consider which tasks to automate and how to enrich remaining human work

Create feedback ecosystems, not just feedback loops – Build mechanisms where humans and AI continuously improve each other's performance

With teams and individuals properly empowered, we now turn to the broader organizational transformation required to scale AI effectively. Chapter 5 examines how to lead and manage organizational change in the age of AI dominance, aligning AI initiatives with company strategy, restructuring processes, and governing AI usage at scale with ethical oversight and agility. The frontier principles that work at team levels must evolve to address enterprise-wide challenges. We'll explore how organizational structures, decision rights, governance models, and leadership approaches must transform to harness AI's full potential while maintaining human judgment at the core of critical decisions.

5

LEADING AI-DRIVEN TRANSFORMATION THROUGH HUMAN-CENTERED CHANGE

AI transformation succeeds when organizations place people at the center of technological change, not technology at the center of organizational change.

The imperative for executives is no longer whether to adopt AI, but how to lead their organizations through the transformation it demands. This challenge requires strategic foresight, cultural sensitivity, and unwavering focus on long-term value creation. Success hinges on recognizing that harnessing AI's potential is as much about organizational strategy and people management as it is about algorithms and data. Implementing AI without a change management strategy is like installing a Ferrari engine in your family sedan without upgrading the brakes or transmission: impressive acceleration followed by spectacular failure at the first curve. Many organizations learn this physics lesson the expensive way.

AI initiatives promise dramatic improvements in efficiency, decision-making, and innovation. Yet without proper alignment and change management, these same initiatives can become expensive experiments that fail to deliver lasting business value. Many companies launch AI pilots or automation projects only to watch them stagnate, unable to translate into sustainable transformation. The difference between isolated experiments and enterprise-wide success lies in effective leadership that balances technological prowess with human sensibility.

This chapter explores how leaders can guide their companies through comprehensive AI transformation. We examine strategic alignment of AI initiatives with overarching organizational goals, ensuring every project serves a clear business purpose. We analyze the necessity of restructuring business processes to integrate AI solutions into daily operations - even the most advanced technology fails when shoehorned into outdated processes.

People remain at the heart of any transformation, and AI is no exception. We explore techniques for managing the human side of change: reducing resistance, building support, and cultivating a workforce that embraces AI as a tool for empowerment. This naturally leads to establishing AI governance frameworks that ensure responsible use aligned with corporate values and regulations.

We then discuss scaling AI solutions across the enterprise, outlining strategies to evolve from one successful pilot to hundreds of deployments while avoiding common pitfalls. Following this, we explain how to develop organizational agility to navigate AI-driven market changes; survival depends on the ability to pivot and adapt as AI continuously shifts competitive dynamics and customer expectations.

Finally, we focus on measuring and sustaining long-term success in AI transformation. Lasting change requires continuous monitoring of impact, reinforcement of new behaviors and processes, and enduring commitment to innovation. By chapter's end, you will have a comprehensive framework for leading an AI-driven change journey from inception to institutionalization.

STRATEGICALLY ALIGNING AI INITIATIVES WITH ORGANIZATIONAL GOALS

Any successful AI transformation starts with a clear linkage between technology initiatives and business strategy.

The excitement around AI was palpable. Across industries, teams were deploying powerful models, automating tasks, and experimenting with new tools. But in one global enterprise, the CTO paused during a leadership retreat and posed a sobering question: "Are we moving with purpose, or just moving?"

It was a moment of reckoning.

A CTO once noted that AI without strategy is like a high-speed ship with no rudder. You're moving with incredible velocity, but mostly you're just doing donuts in the middle of the ocean. It's technically impressive, but the passengers are going to get nauseous, and you aren't getting to port any faster. 'Moving with purpose' prevents your transformation from becoming a very expensive seasickness simulator.

To avoid this, every AI initiative had to be anchored to concrete, prioritized business objectives: enhancing customer experience, reducing operational costs, managing risk, or driving innovation. If a project didn't clearly advance a top-level goal, it didn't belong.

At leading firms, this alignment starts at the top. CEOs and boards don't treat AI as a standalone tech endeavor; they treat it as a business initiative. They ask not just whether a model is cutting-edge, but whether it will drive revenue growth, improve efficiency, or sharpen competitive advantage.

Strategic alignment means weaving AI into the fabric of the company's vision and planning. During annual goal-setting, a retail company might target a 20 percent improvement in customer retention. That goal becomes the foundation for AI-driven personalization and analytics projects.

Research backs this up. High-performing AI organizations are more likely to align their AI strategy with overall business strategy. They're also far more likely to have a clear, enterprise-wide roadmap of AI use cases spanning departments.

Instead of asking, "Where can we use AI?", ask "Could AI help us do this better?" Whether the challenge is high customer churn, slow manual processes, or fraud detection, AI is evaluated as one of several tools and not the default solution.

Consider a bank focused on fraud detection. Its AI models are designed to flag suspicious transactions in real time. A healthcare network aiming to improve patient outcomes invests in predictive systems to prevent readmissions. In both cases, AI is a means to a clearly defined end.

Success also requires clarity on how value will be measured. If the goal is cost reduction, target a 50 percent decrease in processing time or a reduction in labor hours. If the goal is customer experience, track chatbot performance through satisfaction scores or resolution speed. Business metrics and not technical benchmarks define success.

To ensure alignment, cross-functional collaboration is essential. Steering committees that include executives from both business and technology vet AI use cases from multiple angles: Is the solution viable? Does it address a pressing need? A retail company might form a digital transformation council with the CIO, CMO, and business unit heads to align AI investments with growth strategy across marketing, supply chain, and store operations.

Strategic intent must be communicated clearly. When initiatives launch, leaders don't just announce, "We're implementing an AI-powered forecasting tool." They say, "We're improving demand forecasting to reduce inventory costs and better meet customer demand, with AI as an enabler." This framing connects technology to purpose, building understanding and support across the organization.

As market conditions shift, AI portfolios can be recalibrated. A financial services firm facing new regulatory requirements might accelerate AI-driven compliance reporting. Because the roadmap is tied to strategic goals, it remains flexible and responsive.

Most importantly, strategic alignment ensures that technology serves the human and organizational mission. It keeps AI grounded in outcomes that matter to customers, shareholders, and employees. One hospital chain, for example, defined its mission around patient care quality and aligned its AI investments accordingly. It deployed diagnostic assistance tools in radiology and predictive analytics in intensive care units to reduce complications.

This alignment phase lays the foundation. The organization knows which AI projects are happening and why. Vision connects to execution. The next challenge: operationalizing these initiatives by embedding AI into workflows so it can deliver on its promise.

RESTRUCTURING BUSINESS PROCESSES FOR AI INTEGRATION

Aligning AI initiatives with strategy provides direction, but realizing value requires embedding AI into the fabric of day-to-day operations.

It began with a question.

In a quiet conference room, a team of operations managers gathered around a whiteboard filled with boxes and arrows. The diagram traced the familiar steps of their loan approval process, from application intake to final decision. But this time, they weren't just reviewing it. They were reimagining it.

"Which steps in our process," one manager asked, "can be improved or automated with AI?"

The room paused. Then someone pointed to the first box: Loan officer collects applicant information. Another tapped the next: Underwriter manually reviews documentation and credit history.

The process was slow, often taking days. It relied on human judgment, paper trails, and legacy systems. It was ripe for change.

Across industries, AI was no longer a buzzword. It was a tool: precise, scalable, and increasingly indispensable. The team had seen its impact in fraud detection and customer service. Now, they wondered: could it revolutionize loan approvals?

They began with a current-state analysis - a structured approach to AI integration. Every task, decision point, input, and output was mapped. Pain points were surfaced. Then came the future-state vision: a redesigned process where AI transformed more than it assisted.

In this new model, applicants submitted documents online. AI parsed and validated the data in seconds. A credit-scoring algorithm, trained on thousands of historical cases, generated immediate preliminary risk scores and decision recommendations. Low-risk applications were routed for automatic approval, with junior officers performing final reviews. High-risk cases were flagged for deeper human examination.

The cycle time shrank dramatically. What once took days now took hours.

But the transformation was also human. Roles were redefined. Decision thresholds adjusted. Compliance checks updated. Staff retrained on new workflows and tools. The team drew "before" and "after" process maps. Some steps disappeared. New ones emerged like validating AI recommendations or handling AI-human disagreements. The sequence of activities shifted. Controls were redesigned to preserve or improve quality.

This wasn't just happening in banking.

In healthcare, a hospital implementing an AI diagnostic tool for medical imaging faced similar challenges. The redesigned workflow had AI analyze X-rays and MRIs immediately after capture, flagging anomalies for radiologists. The radiologist's role shifted to validating AI findings and focusing on complex cases. New protocols were established: review steps for AI output, fast-track percentages, and safeguards like automatic second reviews for uncertain findings. Diagnosis times dropped. Accuracy held firm. Without restructuring, the AI system would have sat underutilized.

In insurance, AI document analysis tools transformed claims processing. Instead of adjusters manually entering data and performing basic checks, AI triaged documents within minutes. Adjusters received pre-filled case files, allowing them to focus on nuanced judgment calls. The workflow was redesigned to clearly delineate AI-handled tasks (data extraction, simple validations) from human-handled ones (exception handling, final decisions). IT systems were updated to integrate AI outputs into dashboards. Training ensured adjusters could interpret and act on AI-generated insights.

In customer service, AI chatbots handled tier-1 support. But when the bot reached its limits, the process had to enable seamless handoffs to live agents. New protocols guided agents on how to handle chatbot-escalated calls: reviewing interaction summaries rather than starting fresh.

Change leaders knew that piloting AI-infused processes was essential. A retail chain trialed AI-driven inventory management in select stores. Managers discovered that while AI optimized reordering, supplier lead times needed adjustment. These insights were addressed before scaling enterprise-wide. After refining the pilot and documenting best practices, the company saw a 30 percent reduction in stockouts.

Throughout these transformations, one principle held true: frontline employees and middle managers must be involved. Their practical insights into bottlenecks and workflows improved process maps and built buy-in. A tech company implementing AI in software testing engaged QA engineers to help define criteria for AI-generated test scripts. Their participation fostered trust and accelerated adoption.

Policy and compliance considerations surfaced too. In finance, when AI algorithms began approving transactions automatically, risk managers and compliance officers joined redesign discussions. They set autonomy limits and defined audit checkpoints. AI embedded checks directly into workflows.

Ultimately, thoughtfully restructured processes unlocked far greater value from AI investments. Workflows became faster, more efficient, and more adaptable. Employees escaped mundane tasks to focus on higher responsibilities like relationship-building and creative problem-solving. AI's consistency reduced errors and provided rich data for continuous improvement.

But process change creates human impact. Anxiety and resistance are natural. That's why process change and people change are inseparable. The next section addresses managing this human transformation, helping employees understand, accept, and champion new AI-enabled work patterns rather than resist them.

MANAGING THE HUMAN SIDE OF CHANGE – REDUCING RESISTANCE AND BUILDING SUPPORT

Even the most well-aligned, well-designed AI initiative will falter if the people in the organization are not on board.

Human factors often determine transformation success or failure, and AI-driven change is no exception. Employees worry about job impact and managers distrust algorithmic decisions. Research shows a significant percentage of organizational change programs fail to meet objectives, often due to employee resistance or lack of management support. Leading an AI transformation requires proactively addressing these human concerns, turning skeptics into supporters and anxiety into enthusiasm through empathetic, transparent engagement.

Primary fears center on job security when employees suspect AI is sophisticated automation that will render their roles obsolete. This fear has foundation in the broader economy. Leaders within organizations play a crucial role in shaping how these changes are perceived and managed, helping employees focus on emerging opportunities rather than just potential displacement.

When people understand AI tools assist them (taking over drudgery to enable higher-value focus) they're far less likely to perceive threat. Another resistance source is AI trust deficits: employees doubt algorithmic recommendations or fear AI mistakes. This intensifies when AI operates as an unexplained "black box." Additionally, change fatigue sets in if employees have weathered many transformations and view AI as "the next big thing," especially without consultation.

When employees resist AI, they are often signaling a design flaw. Perhaps the AI is ignoring a subtle nuance of the customer relationship, or perhaps the workflow feels dehumanizing. Mine resistance in addition to managing it. It is often the human element trying to correct a drift in the swarm's logic.

Key strategies for reducing resistance and building support:

Clear, Transparent Communication: Leaders must communicate early and often about AI-driven changes, explaining why change is happening, what's changing, and how it affects daily work. Transparency involves acknowledging uncertainties. If role impacts remain unclear, admitting this while committing to share information promptly beats letting rumors fill voids. Throughout transformation, provide progress updates, celebrate early wins ("AI helped us respond to customer inquiries 30 percent faster last month"), and honestly address challenges. This keeps change narrative in leadership's hands while building trust.

Employee Involvement and Ownership: Involve employees in the AI adoption process wherever possible, from seeking input during process redesign to pilot testing with end users and incorporating feedback. When employees have voice (when change happens with them rather than to them) buy-in increases dramatically. A healthcare provider rolling out AI diagnostic aids might form committees of doctors and nurses to evaluate tools in real clinical scenarios and suggest improvements. These clinicians become change champions, having shaped AI implementation in their work. Peers may be more receptive learning from colleagues who've embraced new systems. The more people can experience AI in trials and provide feedback, the more familiarity replaces fear.

Training, Upskilling and Support: Comprehensive training programs help workforces adapt, not just using new AI tools but understanding basic AI concepts to demystify technology. When employees grasp how AI models work and their limitations, mystique diminishes and trust builds. Offer upskilling pathways for those with evolving roles. If certain tasks become automated, provide opportunities to learn new skills and take on new responsibilities, perhaps analyzing AI results, handling exceptions, or focusing on interpersonal job aspects AI cannot perform. This investment sends a clear message: the company wants to evolve with employees, not without them. Provide ongoing support channels: help desks or "AI hotlines" where staff can ask questions or report issues while adapting. Early on, managers should closely monitor employee adjustment, providing additional coaching or temporary assistance when teams struggle. This prevents frustration from festering unnoticed.

Culture of Collaboration and Experimentation: Foster environments where trying new tools and providing feedback is welcomed. Management sets the tone that perfection isn't expected day one, what matters is learning and improving together. When employees see leadership openly addressing AI system problems based on user input, they feel respected and included. Celebrate those leveraging AI effectively to motivate others. If a sales team embraced a new AI CRM assistant and improved results, share their story company-wide. This recognizes adaptability while providing real-life peer examples of how technology improves work. Cross-functional collaboration between AI experts (data scientists, IT) and business users creates environments where AI becomes a team effort. "AI project teams" or liaisons ensure continual two-way communication between technical teams and end users as solutions deploy.

Leadership Engagement and Role Modeling: Employees follow leadership behavioral cues. When executives and managers visibly embrace AI-enabled changes, it send a powerful signal. Use new dashboards or reference AI insights in decision meetings to normalize tools. Leads in team meetings might say, "According to our new AI forecasting tool, demand for Product X is likely to spike next quarter, so we're adjusting plans accordingly." This demonstrates AI trust and encourages reciprocal trust. Identify change champions at various levels that are influential employees excited about AI initiatives. Equip them with knowledge to evangelize benefits to peers. Sometimes colleague or direct supervisor messages resonate more than distant executive pronouncements. Meanwhile, managers should address individual concerns empathetically through one-on-one conversations about personal change impacts. When someone fears role diminishment, managers can discuss how roles might shift and potentially expand in valuable ways as AI handles certain tasks.

Implementing these strategies (i.e., clear communication, employee involvement, training, open culture, and leading by example) significantly reduces resistance. Rather than feeling threatened by AI, employees begin seeing it as work enhancement or growth opportunity. Organizational culture evolves toward tech-forward, change-ready environments where people actively seek AI leverage rather than avoidance.

Positive human adoption critically impacts AI system effectiveness. If employees refuse or begrudgingly use new AI tools, investments won't pay off. Conversely, engaged, supportive workforces find creative value extraction from AI, provide human judgment complementing AI outputs, and spot issues early.

With leadership support and engaged workforce secured, organizations can truly transform. But as employees rely on AI tools and algorithms assume larger decision-making roles, How do we ensure these AI systems operate fairly, ethically, and aligned with our values and policies? The next essential component is establishing robust AI governance and ethical oversight.

ESTABLISHING EFFECTIVE AI GOVERNANCE AND ETHICAL OVERSIGHT

As AI systems become more embedded in critical business processes and decisions, organizations must ensure these systems operate responsibly.

AI governance and ethical oversight create frameworks of policies, practices, and accountability structures guiding how AI develops and operates within organizations. This organizational "rulebook" ensures AI initiatives align with business objectives while adhering to laws, regulations, and company values. It addresses questions such as: Are our AI models fair and free of unacceptable bias? Do we maintain proper data privacy protections? Who bears accountability when AI makes harmful mistakes?

Without clear governance, companies risk deploying AI that creates ethical lapses, reputational damage, or legal violations. Most companies using AI have yet to establish formal governance frameworks, leading to inconsistent practices and elevated risks of data bias or misuse.

AI can amplify human biases or make opaque decisions with significant impact. An AI-powered hiring tool might inadvertently discriminate against certain demographics if trained on biased historical data. Customer service chatbots might generate inappropriate responses. High-profile incidents (e.g., AI chatbots learning toxic language from users, judicial sentencing algorithms showing demographic harshness disparities) highlight how things go wrong without oversight.

Business leaders increasingly recognize that trust and ethics aren't "nice-to-have" AI features but central to adoption. Around 80 percent of business leaders view issues of AI explainability, ethics, bias, or trust as major implementation roadblocks, especially for advanced forms like generative AI. The C-suite understands that without employee, customer, or regulator trust, AI systems never realize potential value. Effective governance builds and maintains that trust.

Key governance components include:

Ethical Principles and Policies: Many companies begin by defining AI ethics principles: commitments to fairness (avoiding discrimination), transparency (explaining AI decisions where possible), accountability (human oversight and responsibility), privacy (protecting personal data), and security. These principles must translate into concrete policies beyond lofty words. Policies might require any AI model making hiring or lending decisions undergo disparate impact testing across gender and ethnic groups before deployment, with periodic post-deployment audits. Another policy might mandate customer notification when interacting with AI versus humans, maintaining transparency. Establishing such policies signals certain standards are non-negotiable in AI projects.

Governance Structure and Roles: Organizations often establish committees or designate roles enforcing AI governance. Common practices include forming AI Ethics Committees or incorporating AI oversight into existing Risk or Compliance committees. These cross-functional groups (IT, data science, legal, compliance, HR) review high-impact AI projects for ethical and risk considerations. Some companies appoint Chief AI Ethics Officers or assign responsibilities to Chief Data Officers. The goal is clear ownership: someone or some group mandated to ask tough questions about AI systems with authority to halt or modify projects failing standards. An AI Ethics Board might review proposals for new machine learning models analyzing employee emails (perhaps gauging engagement or risk) and decide whether privacy implications are acceptable or require additional safeguards.

Standards for Data and Model Management: Good AI governance builds on good data governance. AI models are only as fair and reliable as their training data. Companies should enforce standards for data quality (removing known biases or errors), data privacy (GDPR compliance, controlling sensitive data access), and data security. (The General Data Protection Regulation (GDPR) is a comprehensive data privacy law enacted by the European Union (EU) that came into effect on May 25, 2018. It governs how organizations collect, process, store, and transfer personal data of individuals within the EU and the European Economic Area (EEA)). Models require version control, testing, and validation processes. This might mean establishing AI model inventories: centralized records of all models in use, their purposes, data sources, and performance metrics. Such inventories facilitate periodic behavior reviews. Governance may set rules for model updates and maintenance: predictive models influencing customer outcomes (credit scoring, pricing, medical diagnoses) might require re-evaluation every few months for accuracy and bias, since data patterns shift over time. Integrate these practices into "MLOps" (Machine Learning Operations) pipelines, embedding checks and documentation so compliance becomes workflow-integral, not afterthought.

Regulatory Compliance and External Guidelines: Industries face varying AI regulations or emerging laws. Financial services companies confront model risk management regulations; healthcare providers consider FDA guidance for AI in clinical decision-making; the EU develops a sweeping AI Act regulating systems by risk levels. Effective governance programs track external requirements and ensure AI implementations comply. This could involve conducting impact assessments for high-risk AI applications (similar to GDPR data protection impact assessments) and maintaining documentation on AI system development and testing. Many companies voluntarily align with broader frameworks like OECD AI Principles or industry consortium guidelines, signaling responsible AI commitment. Such alignment provides competitive advantage, building credibility with clients and partners concerned about ethical AI.

The Organization for Economic Co-operation and Development (OECD) is an international organization composed of 38 member countries that promotes policies to improve the economic and social well-being of people around the world.

The OECD AI Principles are the first intergovernmental standard on artificial intelligence, adopted in May 2019 and updated in May 2024. They aim to guide the development and deployment of trustworthy, human-centric AI systems that respect democratic values and human rights.

Core Values of the OECD AI Principles:

Inclusive Growth & Sustainable Development: AI should benefit people and the planet, promoting equity and well-being.

Human Rights & Democratic Values: AI systems must respect privacy, fairness, and freedom, and avoid discrimination.

Transparency & Explainability: AI decisions should be understandable and challengeable by users.

Robustness, Security & Safety: AI systems must be resilient and safe throughout their lifecycle.

Accountability: Developers and operators of AI systems must be responsible for their outcomes.

A bank demonstrating its loan approval AI underwent rigorous fairness testing, with customer hotlines for automated decision appeals, likely engenders greater trust and loyalty.

Governance requires ongoing processes, not one-time setup. Organizations must continuously monitor production AI systems, tracking outcomes for bias signs or rising error rates, gathering user and customer feedback, watching for unintended consequences.

When problems emerge (such as models rejecting unexpectedly high proportions of applications from particular demographics), clear response protocols should exist: possibly retraining models with diverse data, adjusting thresholds, or taking systems offline until fixed. Plan for "what if" scenarios: how to quickly disable or override erratic AI systems.

As AI technology evolves, governance frameworks need updates. Five years ago, few companies had generative AI policies; now many formulate guidelines around AI-generated content vetting and intellectual property issues. Responsive governance programs learn and evolve alongside organizations.

Well-implemented AI governance enables sustainable scaling - it's the difference between a well-paved highway and a Mad Max-style desert chase. Governance creates guardrails within which innovation can flourish safely. Employees and customers are far more likely to embrace AI when they know it's being used thoughtfully.

Ethical oversight and governance build trust both internally (for employees using or impacted by AI) and externally (for customers, regulators, and the public). When organizations show they're not running a "move fast and break things" experiment but rather stewarding a well-governed frontier, people become much more comfortable with AI's expanding role.

With strategic alignment, redesigned processes, supportive people, and robust governance established, foundations are set. Organizations can now focus on scaling AI initiatives from isolated successes to widespread transformation. The next section explores expanding AI solutions across enterprises while maintaining established momentum and control.

SCALING AI SOLUTIONS ACROSS THE ENTERPRISE

Many organizations find themselves trapped in "pilot purgatory," where AI initiatives remain confined to lab experiments or single-department deployments.

To capture AI's full value, it must deploy broadly, touching multiple business facets and driving improvement at scale. Scaling requires not only technical deployment but organizational change on wider scope. 95 percent of organizations struggle to move from pilots to widespread adoption. Avoiding this fate demands deliberate expansion strategy.

Building Scalable Infrastructure and Tools

Ensure technical infrastructure supports enterprise-scale AI. What works for small pilots must be refactored for robust, production-grade environments. This involves investing in cloud platforms or on-premise servers handling large data volumes and concurrent AI workloads. Establishing standardized tools and frameworks prevents teams from reinventing the wheel. Companies might adopt standard machine learning platforms or approved libraries and development practices, enabling models developed across departments to deploy and monitor consistently. Provide experts with common "AI toolkits" and pipelines. About 76 percent of AI high performers have standard AI toolsets company-wide, compared to only 18 percent of other companies. Using shared platforms, models developed for one business unit more easily adapt and scale to other units or geographies, leveraging compatible technology stacks and data architectures.

Governance at Scale

Each deployment should adhere to established principles and policies, which might mean scaling governance processes themselves. AI

Ethics Committees comfortably reviewing five pilot projects may need streamlined processes or delegated oversight facing fifty proposals from different divisions. Organizations address this through tiered approval processes: low-risk AI applications follow standardized checklists for business unit-level approval, while high-impact or sensitive systems escalate for central review. Governance frameworks must scale to avoid bottlenecks. Having governance ingrained from the start means business units know criteria and can self-govern somewhat. Centers of excellence might provide pre-approved tools or documentation templates, easing teams' governance requirement fulfillment while scaling AI use.

Center of Excellence and Cross-Pollination

Many Global 2000 companies establish AI or Analytics Centers of Excellence (CoE) driving scaling. AI CoEs typically comprise central expert teams such as data scientists, engineers, project managers, and "AI translators" mandated to support business units implementing AI solutions, share best practices, and prevent siloed efforts. CoEs act as internal consultants and force multipliers: after one retail division successfully deploys AI demand forecasting, the CoE takes learnings helping other divisions implement for their product lines. They ensure lessons learned in one organizational department communicate to others.

Without such cross-pollination, scaling falters as units start from scratch or repeat mistakes. CoEs maintain repositories of reusable components (i.e., pre-trained models for common tasks, validated datasets) accelerating new projects. They often maintain company "AI playbooks": defined processes taking AI ideas from concept to production that all teams follow. Playbooks outline opportunity identification, business case validation, data acquisition, model development, user acceptance testing, deployment, and post-deployment monitoring. Institutionalizing this ensures scaling doesn't compromise rigor or reinvent processes repeatedly.

Talent and Workforce Scaling

Scaling AI involves scaling human capabilities to use and manage AI. This means significantly enlarging the pool of AI-comfortable employees. Companies might hire more data professionals as model and analysis demand grows. However, more sustainable approaches upskill existing talent enterprise-wide, cultivating analytically savvy staff per department. Organizations run internal AI academies or training programs raising baseline knowledge. Global banks might train hundreds of business analysts and IT staff on machine learning basics and internal AI platform use, enabling them to develop or contribute to AI solutions with less direct central team oversight.

Emphasize continuous learning; they empower broad employee bases to innovate with AI beyond a handful of experts. As skilled employee networks grow, AI ideas and implementations sprout company-wide. Balance remains necessary (complex projects still require expert oversight) but widespread AI literacy greatly aids scaling.

Prioritization and Sequencing

As organizations move from pilots to scale, demand for additional AI use cases quickly become visible. Not all can execute simultaneously, so leaders must prioritize sustaining momentum. Companies typically balance quick wins with foundational projects. Quick wins (automating frequent back-office tasks across departments) show immediate ROI justifying further investment. Complex, foundational projects (overhauling personalization across customer touchpoints) take longer but offer strategic differentiation. Use phased rollouts to roll out simpler use cases to multiple units first while scheduling ambitious, transformative projects in parallel with adequate resources. This creates steady value creation drumbeat.

Multinational healthcare companies might first scale AI scheduling assistants to all clinics (contained use case improving operational efficiency) while parallelly working on grander AI-driven preventive care programs requiring multi-year data consolidation and external partnerships. Prioritizing ensures scaling journeys have value checkpoints rather than long gaps with nothing to show.

Maintaining Quality and Performance

Models working well in one context might need tweaking elsewhere. When predictive models or recommendation engines roll out enterprise-wide, ongoing measurement should confirm similar lift or accuracy achievement in each setting. Local variations (regional customer behavior differences) might require model customization. Organizations should plan for this, not assuming one-size-fits-all without adjustment. Vigilance against model drift and technical debt is crucial; accumulated models and processes can become unwieldy. Investing in robust MLOps capabilities (e.g., automated monitoring, accuracy drop alerts, streamlined model update processes) pays dividends keeping scaled solutions running smoothly. Treat AI systems as products requiring lifecycle management, not one-off projects.

Scaling the Culture Shift

In the early stages of AI scaling, small teams typically lead implementation efforts. But as adoption grows, AI systems begin to impact every part of the organization. At this point, fostering a strong data and AI culture becomes essential. This shift represents enterprise-wide change management. Leaders should continue to share success stories, set clear expectations that using AI tools is now standard practice, and align incentives to support adoption. These signals reinforce that AI transformation is not a temporary initiative but a permanent, organization-wide evolution. When AI is embedded across operations and decision-making it drives significant improvements in productivity, customer satisfaction, and innovation.

Scaling isn't finite but an ongoing journey; as enterprises expand AI use, they must remain agile, ready to adapt to new AI-driven market changes. The next section discusses developing agility and responsiveness needed to maintain pace with rapid AI-driven business environment evolution.

DEVELOPING AGILITY AND RESPONSIVENESS TO AI-DRIVEN MARKET CHANGES

The rise of AI creates ongoing transformation, not one-time events, with market conditions continually evolving around it.

AI innovations are arriving at breakneck speed: what was cutting-edge last year is now commonplace, and competitors rapidly deploy new tools to disrupt entire industries. To thrive, organizations must cultivate agility: the capacity to sense shifts in the environment and respond swiftly and effectively.

What agility requires:

Strategically: Be ready to pivot business models or strategies when AI creates new opportunities or renders existing approaches obsolete.

Operationally: Build the flexibility to reconfigure processes, reallocate resources, and launch AI initiatives faster than ever.

Culturally: Foster a mindset that embraces change as the norm, not the exception - treating disruption as a continuous condition rather than an occasional event.

Sensing and learning early

Agility starts with keeping a pulse on technology and market movements. Leaders should ensure robust monitoring of the external landscape through competitive intelligence, innovation labs, and partnerships with startups and academic institutions. When new techniques emerge (e.g., advances in natural language processing or more capable generative models), agile organizations learn early and quickly assess business relevance.

Often, this means running rapid prototypes or proofs of concept to shorten the time between awareness and action. For instance, forward-looking customer service teams tested automated chat and AI-assisted response drafting as soon as large language models showed promise, while less agile peers waited. Early movers gained hands-on experience and rolled out improvements sooner, capturing customer attention and efficiency gains.

Fast, empowered decision-making

Traditional hierarchies and heavy approval processes slow responses to fast-moving change. Organizations that push decisions closer to the front line (and empower teams to act within clear guardrails) react faster. In practice, this looks like small, cross-functional "SWAT" teams assembled to tackle emerging opportunities or risks.

Consider incumbent banks facing AI-enabled challengers winning younger customers with personalized mobile apps. Agile incumbents rapidly stand up teams from IT, marketing, and product to add AI-driven personalization without waiting for the annual planning cycle. Working in sprints, they deliver weekly increments. Decentralized decisions and structured experimentation help avoid being caught flat-footed.

Designing for modularity

Just as modular technology (APIs, microservices) accelerates upgrades and replacement, modular organizations enable rapid reconfiguration. Cross-trained talent and broad skill sets let people shift roles as priorities change. Data science teams, for example, can form flexible "swarms" that focus on urgent projects, optimizing supply chain models one quarter, then driving marketing analytics the next. Contracts with external partners and cloud providers should also be structured for rapid scaling and portability to avoid capacity bottlenecks that undermine agility.

Continuous improvement powered by AI

Agile organizations treat AI not as a one-off automation but as a continuous learning engine. Retailers using AI for inventory management, for example, feed outcomes (stockouts, overstocks) back into models and processes to refine rules dynamically as trends or seasons shift. Real-time data and insights make this feasible: AI-driven dashboards and KPI monitors can flag emerging issues or opportunities, shrinking feedback loops.

In healthcare, if AI detects a sudden rise in appointment no-shows at a clinic (perhaps due to local transport disruptions or a contagious outbreak) agile management can immediately pilot mitigations such as overbooking or telehealth, rather than discovering the problem in a quarterly review.

Leading for change

When leaders openly explore possible strategic pivots (e.g., "What if AI enables new service models that require us to shift our product mix?"), they signal that change is an opportunity, not a threat. Many organizations codify this with scenario planning: preparing responses if competitors achieve specific AI milestones or if regulations constrain current approaches. This preparation, both mental and practical, accelerates real-world responses.

Celebrating quick, well-judged responses to AI-driven market shifts builds confidence and psychological safety. When marketing, for instance, adapts a campaign within days because AI analytics reveal a sudden sentiment change, leaders should highlight the win. Over time, people act decisively because they trust they won't be penalized for prudent, fast decisions amid uncertainty.

Agility with discipline

Agility doesn't mean chasing every trend. The essential capability is to rapidly assess changes and decide whether to pivot or double down. In a world where breakthroughs in algorithms or surges in IoT data can reshape competition overnight, the ability to iterate on strategy becomes as important as the strategy itself. Many organizations are shifting from rigid annual plans to rolling 90-day strategic reviews. These checkpoints don't imply changing direction every quarter; they ensure targets and initiatives adjust as new information emerges.

Agility is foundational to future-proofing in an AI-driven world. Agile companies absorb shocks and seize opportunities faster than slower rivals. As AI raises the bar through hyper-personalized experiences and real-time optimization, agile organizations adapt smoothly, avoiding the friction of clumsy transitions. This adaptability reinforces every facet of AI transformation: aligning strategy, redesigning processes, engaging people, governing responsibly, and scaling effectively. It ensures organizations keep evolving.

With an agile, change-ready culture in place, the final step is to ensure the benefits of AI transformation are measured and sustained over time. The concluding section will examine how to evaluate AI outcomes and maintain momentum well into the future.

MEASURING AND SUSTAINING LONG-TERM SUCCESS IN AI TRANSFORMATION

The final step of any major change initiative is to measure its success and ensure the improvements endure. Transformation often suffers from short memory: intense excitement at launch, then amnesia a quarter later when the next big thing arrives. Measurement is the breadcrumb trail through the forest of organizational attention: it reminds everyone where we started, why we invested, and what value we're realizing. Once AI projects are live and systems embedded, leaders must step back and ask two questions: Did we achieve the outcomes we set out to deliver? and How do we sustain momentum over time?

Answering these questions isn't a one-off review. It requires an operating rhythm that includes:

- Clear success metrics tied to strategic objectives
- Regular performance reviews to track progress and surface gaps
- Adaptive adjustments to sustain gains while fueling further innovation

Done well, this ongoing evaluation turns AI transformation into durable value, not a short-lived spike.

Defining the Right Metrics

AI's impact is multifaceted, so success must be measured across dimensions. Every initiative should be anchored to specific business KPIs from the outset. For example:

- Revenue growth targets (e.g., increase annual sales by X percent)
- Cycle-time reductions (e.g., halve average processing times)
- Quality improvements (e.g., reduce error rates)
- Customer outcomes (e.g., raise CSAT/NPS)

These become the primary outcome metrics.

Example: If supply-chain optimization targets a 15 percent reduction in inventory carrying costs, finance should monitor the relevant metrics quarterly to verify materialized savings.

Beyond direct outcomes, track leading indicators that explain why results are (or aren't) happening:

- Adoption & usage — Are employees and customers using the AI tools?
- User satisfaction — Are tools intuitive, useful, and trusted?
- Operational resilience — Are we better at handling disruptions?
- Innovation velocity — Are we experimenting and shipping faster?
- Cultural alignment — Are teams leaning into data-driven decisions?

Adoption and Usage

If people don't use the tools, value won't show up. Monitor:

- Login and active-use rates to gauge engagement
- AI-handled vs. manual decisions/transactions to quantify automation uptake
- User satisfaction and task-level feedback to spot usability, trust, or workflow issues

Low adoption is a signal often pointing to training gaps, UX friction, or change-management needs. Tracking these metrics regularly helps remove barriers early.

Performance and Accuracy

Ensure the systems do what they're supposed to do reliably and well:

- Model performance (accuracy, precision/recall, error rates, drift)
- Reliability (system uptime, latency)
- Task-specific KPIs (e.g., for chatbots:
- Resolution rate: percent of inquiries solved without human escalation
- Response time: how quickly users receive help)

Continuous monitoring enables timely retraining, tuning, and process adjustments before issues erode trust.

Business Outcomes and ROI

Ultimately, AI should move the big needles:

- Revenue growth, profit margins, market share
- Operational cost ratios and productivity
- Customer retention and satisfaction

Create benefits-realization dashboards that consolidate KPIs across initiatives to show how AI contributes to strategic goals, making progress visible, value communicable, and investment decisions clearer.

If the company invests $10M in AI, leaders should ask, What tangible returns or strategic advantages did we realize? Examples include:

- Net new customers acquired
- Improved safety and compliance outcomes
- Cost savings and efficiency gains

Not all benefits are directly financial. Improvements in decision quality, time-to-insight, or product development speed can be assessed via structured stakeholder feedback, internal case studies, and proxy metrics (e.g., time-to-decision, release cadence). Let data and not gut feel tell the story.

Integrate AI metrics into monthly ops reviews and quarterly business reviews, not as standalone "AI stats" but as part of core business reporting. This signals that AI is a driver of performance, not a side project.

Continuous Improvement and Iteration

Measurement should fuel learning, not just self-congratulation. Treat misses as feedback loops:

- If model accuracy slips, investigate data drift or new scenarios and retrain or re-feature.
- If a process improvement stalls at 10 percent vs. a 20 percent target, explore training upgrades, workflow redesign, or system tuning.

Formalize this with post-implementation reviews at 3, 6, and 12 months, assessing outcomes against targets and documenting lessons learned. Feed insights into the next wave of projects and model refreshes. Mature organizations evolve into a "sustain and improve" mode: small, agile teams rotate through functions to tune models, UX, and processes continuously (akin to Lean/Six Sigma continuous improvement).

Sustaining the Change

Human systems tend to snap back to the familiar, especially through leadership changes or shifting priorities. Lock in gains by embedding changes into structures and culture:

Institutionalize New Processes: Update SOPs, policies, and job descriptions to reflect AI-enabled work. If meetings now start with a data dashboard review, make it a formal step. If approvals are automated, remove the manual backdoors.

Maintain Leadership Focus: Keep AI outcomes on strategy and performance agendas. Ensure roles like CDO/CAIO are empowered. Leaders should routinely ask about AI's contribution and recognize teams that deliver and sustain value.

Refresh Skills and Talent: Offer ongoing training, onboard new hires into AI-enabled practices, and consider annual AI bootcamps or digital upskilling. This prevents skill atrophy and retains talent that values continuous learning.

Celebrate and Reinforce Successes: Revisit early use cases after 12–24 months to showcase measurable outcomes. Pair the data with customer or employee stories. Recognize not just launch teams but those who continuously improved solutions to help combat change fatigue.

Adapt Strategy as Needed: Reassess alignment as business goals evolve. Scale down initiatives that have delivered their value; scale up those addressing new priorities. As capabilities mature, expect a shift from foundational builds to edge innovation to stay flexible.

By rigorously measuring outcomes and embedding gains into the operating model, AI transformation becomes a continuous evolution. Organizations discover new applications, track what works, retire what doesn't, and keep people, processes, and strategy adaptable. In short, they become learning organizations powered by data and AI, constantly refining themselves to stay ahead.

SUMMARY

In this chapter, we've traced the full arc of leading and managing organizational change in an era increasingly shaped by AI, from setting the right strategy and redesigning processes to empowering people, upholding ethics, scaling effectively, staying agile, and ultimately securing long-term success.

AI transformation is not a sequence of technology deployments; it's a holistic organizational journey. It requires human insight, cultural alignment, and deliberate leadership. The principles outlined here are designed to help you lead this transformation in your enterprise, balancing AI's vast potential with the structure and discipline needed to turn possibility into sustained progress.

Chapter takeaways:

- AI transformation is about organizational change, not just technology. Success hinges on aligning technology with business strategy, redesigning processes, and managing the human side of change.
- Strategic alignment is non-negotiable. Every AI initiative must clearly advance business objectives. Companies that integrate AI strategy with overall business strategy achieve far greater impact.
- Process transformation, not just acceleration. AI's real value lies in reimagining how work gets done, not simply speeding up existing workflows. This requires thoughtful redesign with people at the center.
- The human factor determines success or failure. Overcoming resistance, building trust through transparency, and providing robust training are essential. Employees must view AI as an enabler, not a threat.
- Governance provides guardrails for innovation. Strong ethical oversight and clear accountability build trust and enable sustainable scaling, empowering innovation rather than constraining it.
- Scaling requires standardization and knowledge sharing. Common platforms, tools, and Centers of Excellence help successful pilots scale enterprise-wide. Governance must evolve alongside deployment to prevent bottlenecks.
- Agility is the new competitive advantage. Organizations must sense market shifts quickly and respond effectively. This calls for flexible structures, empowered teams, and a culture that embraces continuous adaptation.
- Measurement drives sustained success. Multi-dimensional metrics, regular reviews, and iterative improvements ensure AI benefits endure. Metrics should integrate into core business reporting, not sit in isolation.

- Institutionalize change to make it stick. Update processes, maintain leadership focus, refresh skills continuously, and celebrate wins to prevent regression to old habits.
- AI transformation is a journey, not a destination. Organizations must evolve as AI capabilities and market conditions shift. The ultimate goal: becoming a learning organization powered by data and AI.

The next chapter explores real-world examples and emerging trends, including how these principles converge in practice and what patterns leaders should anticipate as AI continues to evolve and redefine business landscapes.

6

INDUSTRY SCENARIOS AND FUTURE TRENDS

AI transformation is happening now across every major industry, each grappling with unique challenges and opportunities.

The patterns of success and failure emerging from these early adopters provide a roadmap for organizations still charting their course. While the technology may be universal, its implementation is anything but.

Roughly seven in ten organizations report using AI in at least one business function, yet most still struggle to turn pilots into scaled impact. Recent research shows only a minority achieve significant financial benefits, with the main blockers rooted in people, process, data, and operating-model challenges, not the algorithms themselves.

The following seven industry scenarios are composite narratives drawn from real organizations navigating the complexities of AI adoption. These examples are designed to do two things:

- Illustrate core change management principles from earlier chapters in a variety of organizational contexts.
- Surface nuances that industry-specific pressures and priorities introduce.

Here's what's ahead:

Financial Services: A global bank wrestles with risk, compliance, and employee skepticism as it introduces AI into fraud detection and customer service.

Healthcare: A hospital network implements diagnostic AI, raising questions about clinical judgment, liability, and bedside manner.

IT & Software Development: A software firm adopts AI pair-programming tools, only to discover that even engineers need change management.

Retail: A big-box retailer turns to AI for demand forecasting and personalization, navigating union concerns and frontline adoption.

Manufacturing & Supply Chain: An industrial firm deploys AI-driven robotics and predictive maintenance, redefining shop-floor labor and trust.

Public Sector: A city government integrates AI into public services, balancing efficiency with citizen transparency and democratic oversight.

Future Trends: A speculative look at an organization five years from now, where AI is ubiquitous, invisible, and deeply entwined with strategy, operations, and ethics.

Across these scenarios, some patterns hold:

- Success hinges on executive clarity, employee involvement, and ethical foresight.
- Resistance is natural, but not immovable.

- Culture is not a barrier to overcome; it's the medium through which change must travel.

BANKING ON INTELLIGENCE: THE APEX BANK CASE (A HYPOTHETICAL EXAMPLE)

When customers demand 24/7 service but regulatory constraints limit innovation, how does a traditional financial institution modernize without losing its foundation of trust?

The Challenge

By early 2024, Apex Bank was underwater on customer service. Inquiry volumes had tripled in five years while satisfaction steadily fell. Customers now expected instant, 24/7 help, but the legacy, phone-centric model couldn't scale economically. Peak-hour waits stretched past 45 minutes, and 60 percent of traffic was routine "how-do-I" questions burning scarce, expensive agent time.

New CEO Sarah Waters didn't mince words: "We had a choice: reinvent service around the customer or watch them defect to digital-first rivals already delivering the experience we couldn't."

Leadership agreed AI could absorb repetitive volume. But in banking, risk and trust are non-negotiable: every interaction needs an audit trail, privacy must be preserved end-to-end, and any wrong answer risks regulatory attention, brand damage, and customer churn.

The Strategic Response

Waters framed the program not as an IT project but as a customer-experience transformation, modeled on proven patterns from leaders like Erica/Cora/Ceba:

Human–AI Partnership, by Design: The virtual assistant would contain routine inquiries and triage, while complex, high-value issues flowed to skilled humans. The aim: elevate human work, not replace it.

Regulatory-First Architecture: Guardrails were built in from day one: no advice beyond approved scripts, clear AI disclosure, verbatim interaction logs, explanations on demand, and confidence-based escalation. (Think model risk management aligned to banking standards, with red-teaming and bias/accuracy testing before each release.)

Change Management as a Core Workstream: Apex treated employees as co-designers. Veteran agents became “AI Supervisors”: tuning content, monitoring quality, and deciding when to retrain models so the system learned with the workforce, not at their expense.

Implementation Strategy

Governance Foundation (Months 1–2): A cross-functional AI Steering Committee (customer service, compliance, risk, infosec, legal, and CX) codified rules of the road: use-case boundaries, PII handling, explanation standards, human-in-the-loop thresholds, and incident response. Templates for model cards, prompt libraries, and evaluation playbooks set the tone for safe scaling.

Pilot Build (Months 3–5): The team targeted the “boringly valuable” top-50 intents: balances, transactions, card status, branch hours, dispute status, password resets. Answers were grounded on an approved knowledge base (no free-form web), with confidence gating and one-tap human escalation baked into every channel (app, web chat, and IVR deflection).

Workforce Readiness (Months 4–6): Agents trained on AI-assisted workflows: picking up escalations with full conversation history, leveraging auto-summaries, and using next-best-action prompts. Senior reps stepped into AI Supervisor roles, tuning prompts, tagging bad outputs, and approving content updates. Incentives shifted to quality and resolution, not handle time alone.

Controlled Launch (Months 6–8): A staggered rollout (off-peak windows, opt-in cohorts, and A/B tests) let supervisors watch interactions live and "break-glass" to a human instantly if needed. Weekly "quality councils" reviewed transcripts, complaints, and false-positive/negative patterns; releases only shipped after hitting defect and fairness thresholds.

Managing Resistance:

Employees (Job Security): The bank published a no-layoff commitment tied to the program's first year and showed, in side-by-side demos, that AI would remove the drudgery (PIN resets, balance checks) and free reps for diagnosis, empathy, and retention. Promotions into specialist and supervisor roles made the upside tangible.

Regulators (Model Risk & Conduct): Apex brought supervisors into pre-pilot reviews with regulators, sharing governance artifacts (interaction logs, change controls, red-team results). That transparency earned early approvals while rivals were still revising paperwork.

Customers (Trust & Choice): The assistant introduced itself as AI, offered a human in one click/keyword, and showed a "source of truth" link when answers were policy-sensitive. That combination of speed plus honesty diffused skepticism.

Critical Incident and What Changed

Two months in, the assistant gave a customer an out-of-date promotional APR for a loan product. The error was caught within hours via a low-confidence spike alert and a supervisor spot check. Apex activated its crisis protocol:

Immediate Outreach: Correct terms, a waiver of fees, and a personal apology from the branch manager.

Root-Cause Post-Mortem: The response referenced a stale content variant; procedures changed so time-boxed promos auto-expire and confidence thresholds tighten for rate/fee topics.

Learning Loop: Mandatory refresher for agents and supervisors on AI limitations and manual override standards.

Counterintuitively, customer trust rose in the next NPS wave: people saw that humans were in control, and mistakes triggered rapid, transparent corrections, a hallmark of mature programs at banks like Erica/Cora.

Results After 18 Months

Operational Impact

- 65 percent containment of routine inquiries (AI resolved without human handoff)
- Average wait time down from 12 minutes to 4 minutes (peak spikes tamed)
- Service capacity +40 percent with no net headcount increase
- Cost-to-serve −25 percent across assisted channels

Customer Experience

- CSAT +15 points; complaints about hold times down >50 percent
- 24×7 coverage with consistent quality and clear human-escalation paths
- Better first-contact resolution as agents focus on complex journeys (disputes, hardship, onboarding)

Employee Outcomes

- Zero involuntary reductions in customer service

- ~30 percent of reps advanced into AI Supervisor or Customer Success roles
- Employee engagement up as work shifted from rote tasks to diagnosis and relationship-building

Why It Worked (and Scales)

- Start small, scale what sticks: Target high-volume, low-risk intents first; keep humans in the loop until the data proves quality.
- Compliance as accelerator, not brake: Treat governance, logging, and explainability as product features that enable approvals and trust.
- Learn with your people: Make frontline experts owners of quality (prompt tuning, content curation, escalation policy).
- Confidence-aware design: Grounded answers, confidence thresholds, and instant human access; the same patterns behind successful bank assistants globally.

Questions for Reflection

- How might Apex's governance-first approach apply to AI implementations in other regulated industries?
- What role should existing employees play in designing AI systems that will change their work?
- When AI makes mistakes in high-trust environments, how should organizations balance transparency with reputation protection?
- How can traditional organizations use regulatory requirements as innovation accelerators rather than constraints?

HEALING WITH ALGORITHMS: THE MERCY HEALTH CASE (A HYPOTHETICAL EXAMPLE)

When radiologist shortages meet soaring demand for medical imaging, can AI fill the gap without compromising patient care or physician autonomy?

The Challenge

Dr. Jennifer Walsh faced a near-impossible equation. As Chief Medical Officer at Mercy Health System - a regional network serving 2.3 million patients across four states - she watched imaging volumes climb 18 percent annually while the radiology department grappled with chronic understaffing. The impact was stark: patients waited up to three weeks for scan results, delaying critical cancer treatments and cardiac interventions.

"We had radiologists reading 400 studies per day when best practice suggests around 200," Walsh recalls. "Everyone was burning out, and patient care was suffering. We needed a way to scale expertise without sacrificing quality."

Hiring more radiologists wasn't realistic. A nationwide shortage pushed salaries higher and made geographic flexibility a must - something Mercy couldn't always accommodate. Meanwhile, emergency departments demanded faster reads, oncologists needed immediate staging results, and patient satisfaction scores reflected growing frustration with delays.

Unlike customer-service chatbots that can apologize for mistakes, medical AI errors carry life-and-death consequences. Any solution had to win physician trust, meet regulatory standards, and integrate safely into clinical workflows that prioritize patient safety over speed.

The Strategic Response

Walsh's breakthrough drew on her surgical background: physicians trust tools endorsed by respected peers, not edicts from above. Rather than framing AI as an efficiency play, she asked Dr. Robert Smith, Mercy's most experienced thoracic radiologist, to lead the evaluation.

"My first reaction was defensive," Smith admits. "I'd spent thirty years honing pattern recognition. The idea that an algorithm could match that felt diminishing."

Scientific discipline won out. Smith agreed to test AI systems objectively, as if running a clinical trial.

After rigorous testing such as reviewing validation studies, running historical case challenges, and demanding transparency into model reasoning, Smith endorsed AI assistance. His credibility mattered more than any vendor demo. Mercy selected a system specializing in chest imaging, with strength in detecting lung nodules suggestive of early-stage cancer.

Implementation Strategy

Mercy avoided a false binary between human and machine. For six months, every chest CT received dual interpretation: a radiologist read and an AI analysis. The approach validated accuracy, built confidence, and protected standards during transition.

AI identified suspicious nodules in 12 percent more cases than human readers alone, particularly subtle early-stage findings. False positives were initially higher, but the oversight committee judged that acceptable: missing cancer is far costlier than investigating a false alarm.

The AI analyzed scans immediately post-acquisition and produced preliminary reports with confidence scores. Cases were triaged by urgency: likely-normal studies with high AI confidence were dispatched quickly; flagged cases received deeper attention. Radiologists spent less time on routine normals and more on complex cases: the work where their expertise matters most.

Managing Resistance

Dr. Sarah Kim was a radiologist with 25 years of experience that initially declined AI assistance, relying on her proven judgment. Her conversion, driven by clinical evidence, became a turning point.

Three months into the pilot, Kim reviewed a routine screening in a 45-year-old executive showing what appeared to be a benign nodule. Her initial read suggested 12-month follow-up. The AI flagged the area with high concern. Protocol required investigation. Additional imaging revealed features consistent with early adenocarcinoma; biopsy confirmed malignancy, caught nearly a year earlier than traditional surveillance would have.

"That case changed everything," Kim reflected. "The AI didn't replace my judgment, it made me look harder. The patient received curative surgery instead of facing advanced disease."

Surprisingly, some newer physicians were more anxious than veterans, concerned that AI would atrophy their diagnostic skills. Mercy responded with "diagnostic maintenance" sessions (scheduled AI-free reads) and explanatory model outputs that turned AI into a teaching tool.

Critical Incident and Learning

Four months in, the AI flagged a routine pediatric chest X-ray as "high concern for pneumonia," despite no symptoms. The ED physician, trusting the alert, prescribed unnecessary antibiotics and ordered extra imaging.

Root-cause analysis showed the model was trained primarily on adult data, reducing pediatric reliability. Mercy didn't bury the mistake or abandon the system. They instituted age-specific confidence thresholds, required human verification for all pediatric flags, and updated training to emphasize limitations and the primacy of clinical judgment.

"That mistake reminded us AI is a tool, not a substitute for thinking," says Dr. Martinez. "We got better by learning when to trust the AI and when to trust ourselves."

Results and Impact (18 Months)

Clinical outcomes

- +23 percent increase in early-stage cancer detection
- Turnaround time cut from 2.8 days to ~6 hours
- +15 percent improvement in diagnostic accuracy for subtle nodules
- +28 percent rise in patient satisfaction with imaging services
- Operational impact
- +35 percent radiologist productivity without quality degradation
- −60 percent ED wait times for imaging results
- Significant reductions in burnout indicators
- No staff reductions despite higher throughput

System-wide effects

Success in radiology opened doors for AI in cardiology and pathology. Physician confidence rose across departments, and Mercy's regulatory posture strengthened through proactive, safety-first governance.

Why it worked

- Peer-led credibility to earn adoption

- Safety-first design (parallel reads, targeted triage, explicit thresholds)
- Workflow-native integration over model flashiness
- Transparent learning culture that turns incidents into improvements

Questions for Reflection

- How can organizations identify and leverage clinical champions to drive AI adoption in professional environments?
- What role should parallel implementation play when AI systems affect critical outcomes?
- How should organizations balance AI efficiency gains with the need to maintain human expertise?
- When AI makes errors in high-stakes environments, how can organizations turn mistakes into learning opportunities rather than reasons for abandonment?

CODE MEETS CODER: THE DEVINNOVATE CASE (A HYPOTHETICAL EXAMPLE)

When software developers who build AI systems for clients must adopt AI tools themselves, how do you convince skeptical programmers to embrace automation of their own craft?

The Challenge

In late 2023, DevInnovate Inc. (a 500-person software consultancy) found itself in a peculiar bind. For five years the firm had built AI solutions for Fortune 500 clients, yet its own development teams remained stubbornly manual. While they automated supply chains and customer service for others, DevInnovate's programmers wrote every line of code by hand.

CTO Sarah Kim saw the competitive risk: "Clients were asking why our development cycles took months when AI-powered teams elsewhere delivered in weeks."

The market was moving quickly. Industry surveys showed roughly 70–80 percent of developers were already using or planning to use AI coding assistants. DevInnovate's engineers resisted, not out of technophobia, but pride in craftsmanship. Coding was their art form. To many, AI assistance felt like suggesting that Beethoven needed help composing symphonies.

Technical lead Marcus McClure voiced what others were thinking: "I've spent 15 years mastering architecture. Why would I trust an algorithm to write code I can do better myself?"

Kim called it the craftsperson's dilemma: how to automate parts of expert work without devaluing the expert.

The Strategic Response

Kim reframed AI as a career imperative rather than a corporate cost play: "The question isn't whether AI will change development; it's whether you'll be the developer using AI or the one replaced by those who do."

DevInnovate launched an optional pilot. Twenty developers across product lines volunteered for a three-month test, deliberately mixing enthusiasts and skeptics, senior architects and juniors. The approach avoided the backlash that mandates can trigger and created credible peer advocates instead of reluctant compliance.

Implementation Strategy

The assistant was embedded directly into familiar IDEs like Visual Studio Code: no new tools, no forced process change. Developers could accept, edit, or ignore suggestions, maintaining full control.

Unexpectedly, senior engineers became some of the most enthusiastic users. The assistant chewed through boilerplate, test scaffolding, and documentation, precisely the tasks veterans least enjoyed.

"I thought it would replace creative work," one said. "Instead, it freed me from the grunt work I've done for 20 years."

Training emphasized hands-on practice. Teams tackled coding challenges with and without AI to establish where it excelled (pattern recognition, syntax accuracy, rapid prototyping) and where human judgment remained paramount (architecture, domain logic). Developers began treating AI as a powerful autocomplete, not an architect.

Managing Resistance

Early adopters' sprint metrics improved visibly. Demos showed AI-assisted teams shipping features while manual teams were still debugging. Skeptics noticed colleagues spending less time on repetitive implementation and more on design and problem-solving.

McClure's turn came during a complex API integration: "I was writing dozens of similar endpoint handlers. The AI generated them in minutes, so I could focus on the business logic that mattered."

New responsibilities emerged organically. Some developers became prompt specialists, mastering techniques to coax higher-quality suggestions. Others curated internal examples so the assistant reflected DevInnovate's idioms and patterns. The roles signaled career growth through AI collaboration, not despite it.

Critical Incident and Learning

Two months in, a junior developer committed AI-assisted code that included copyrighted snippets from an open-source project with restrictive licensing. Legal flagged the risk and escalated to leadership.

DevInnovate responded by integrating license-compliance scanners, adding attribution guidelines requiring developers to document AI contributions, and instituting reviews to ensure AI-generated code met the same quality and legal standards as human-written code.

"That incident taught us AI introduces new risks," Kim said. "We needed guardrails to protect both quality and our legal standing."

The transparent response increased developer confidence: leadership was serious about responsible use without abandoning the tools.

Results and Impact (6 Months)

Development Metrics

- +30 percent development velocity
- Bug rates down, with syntax issues caught earlier in the IDE
- Documentation quality up, aided by AI-generated comments and summaries
- Time-to-deployment cut by ~2 weeks per project on average

Developer Experience

- Higher job satisfaction in pulse surveys
- More time spent on architecture and problem-solving
- Internal innovation projects increased as teams gained capacity

Business Impact

- Faster client delivery while maintaining quality
- Improved project profitability from reduced engineering hours
- Stronger positioning in RFPs and client briefings
- Better retention, as work shifted away from repetitive tasks

Why It Worked

- Clear personal framing: AI as a career-critical capability, not a threat to craftsmanship
- Voluntary pilot that produced visible wins and credible peer advocacy
- Workflow-native integration in existing IDEs and pipelines
- Governance and guardrails that addressed IP, quality, and compliance without stalling momentum

Questions for Reflection

- How can organizations introduce AI tools to highly skilled professionals without threatening their sense of expertise and value?
- What role should voluntary adoption play when implementing AI tools that fundamentally change how work gets done?
- How do you balance the efficiency gains of AI assistance with the need to maintain human skill development?
- When AI tools create new risks (like IP violations), how should organizations respond without abandoning beneficial technologies?

RETAIL REDUX: THE BRIGHTMART CASE (A HYPOTHETICAL EXAMPLE)

When customer expectations demand perfect inventory while profit margins demand efficiency, can AI solve retail's oldest problem without alienating frontline teams?

The Challenge

BrightMart's 847 stores across North America faced mounting pressure in 2023. Empty shelves cost an estimated 8 percent in lost sales, while excess inventory tied up $340 million in working capital. Store managers relied on gut instinct, spreadsheets, and last year's patterns - methods that worked when customer behavior was predictable.

"We had managers who knew their communities intimately but couldn't scale that knowledge," recalls COO Lisa Ying. "Store 445 in suburban Phoenix nailed seasonal items because the manager understood local demographics. Meanwhile, Store 12 in rural Montana struggled with the same products because their manager was guessing."

Managers prided themselves on knowing local nuances such as the timing of back-to-school rushes, weather-driven umbrella sales, and demand spikes from community events. Any AI solution had to respect this expertise while providing analytical power beyond human reach. Traditional forecasting models based on historical averages failed to capture these dynamics. BrightMart needed machine learning that could process weather data, economic indicators, local events, and demographics, as well as present insights in a way managers could trust.

The Strategic Response

Ying framed the initiative as empowerment: "We're not automating your decisions; we're giving you superpowers to see around corners."

Regional VP Mark Rodriguez, a former store manager, became the program's champion. His credibility with frontline teams made adoption conversations authentic: "Mark could walk into any store and speak their language," Ying notes. "When he endorsed AI forecasting, it carried weight no executive deck could."

BrightMart launched a voluntary pilot with 30 managers across diverse markets. This avoided backlash and created advocates based on real results rather than compliance.

Implementation Strategy

The AI dashboard displayed recommended order quantities, confidence levels, and plain-language explanations. Managers could override recommendations with a click, and the system learned from those adjustments.

"We could have shown complex probability charts," says IT Director Chris Hetzer. "Instead, we focused on actionable insights managers could use immediately."

Training combined short digital modules with peer-led sessions. The most persuasive moments came from success stories. Patricia Santos from Phoenix became an evangelist after the system predicted a spike in patio furniture sales three days before an unexpected heat wave: "We sold out while competitors still had winter inventory on display."

Managing Resistance

Veteran manager Frank Price initially dismissed algorithmic advice: "I know my customers better than any computer."

In month three, AI recommended a 40 percent increase in snow shovels for his Denver store, weeks before he expected winter demand. He decided to follow the AI recommendation while noting in the system that he expected it to fail. An early October snowstorm hit, and his store cleared $47,000 in extra revenue while competitors faced stockouts.

"That's when I realized the machine was seeing patterns I couldn't," Price admits.

Regional concerns varied: rural managers feared urban bias, suburban managers doubted demographic sensitivity, and urban managers questioned adaptability to rapid shifts. BrightMart countered with transparency, showing how the model incorporated local weather, school calendars, economic signals, and event schedules.

Critical Incident and Learning

Five months in, the AI recommended a surge in grilling supplies for Minneapolis stores during early March cold. Most managers ignored it. Jennifer Walsh trusted the system. The AI had processed long-range forecasts predicting a warm spell. When temperatures hit 75°F mid-March, Walsh's store dominated the market while competitors scrambled.

"The AI was looking at meteorological data I didn't even know existed," Walsh says.

Results and Impact (12 Months)

Inventory Performance

- Stockouts down 31 percent, excess inventory down 23 percent
- Gross margin +2.8 percent through better demand matching
- Inventory turnover improved from 6.2x to 8.1x
- Working capital reduced by $89M

Manager Experience

- Ordering time down 35 percent, freeing time for customer engagement
- Confidence in decisions up significantly
- Job satisfaction improved as emergency restocking declined
- Regional performance gaps narrowed

Customer Impact

- Product availability scores up 18 percent
- Stockout-related complaints down 42 percent
- Loyalty metrics showed steady improvement

Why It Worked

- Positioned AI as augmenting expertise, not replacing it
- Voluntary pilot created authentic advocates
- Simple, transparent interface built trust
- Local data integration respected manager knowledge

Questions for Reflection

- How can organizations implement AI systems that enhance rather than threaten the expertise of experienced employees?
- What role should voluntary participation play when rolling out AI tools that fundamentally change how work gets done?
- How do you balance the need for algorithmic consistency with respect for local knowledge and decision-making autonomy?
- When AI systems outperform human intuition, how should organizations communicate these successes without diminishing employee confidence?

MANUFACTURING'S NEXT ACT: THE GLOBALPARTS CORP CASE (A HYPOTHETICAL EXAMPLE)

When a century-old manufacturer decides to become a "smart factory," how do you convince seasoned workers that AI-powered machines are allies, not adversaries?

The Challenge

GlobalParts Corp's Plant 7 in Columbus, Ohio embodied industrial America's contradictions. Built in 1952, the facility produced precision components for aerospace and automotive clients using a mix of cutting-edge CNC systems and equipment older than many employees. By 2023, unplanned downtime was costing $2.3 million annually, while global competitors offered shorter lead times and lower prices.

Plant Manager Tom Rodriguez recalls: "We had machines failing without warning, rushing emergency repairs that cost triple the normal rate. Meanwhile, our workforce with on average thirty-plus years of experience could often predict failures days before they happened just by listening to engine sounds or feeling vibrations."

Human expertise that could save millions was locked in individual knowledge, unavailable when those experts weren't on shift. Night supervisor Maria Santos could detect bearing problems by sound, but her insights vanished when she went home. Day shift leader Matt Brown knew which pumps failed during summer heat, but weekend crews didn't.

GlobalParts turned to AI-powered predictive maintenance combined with supply chain optimization to capture and scale this expertise. Success depended on convincing workers that machines learning from their knowledge meant partnership, not replacement.

The Strategic Response

Rodriguez acknowledged the workforce's skill before introducing technology: "You've been our AI for decades. Now we're teaching machines to learn what you know."

Union partnership shaped the rollout. Local 485's president worried about job loss. Rodriguez countered: "We need your members to teach the machines. Without their knowledge, this system fails."

The agreement guaranteed no layoffs due to AI and retraining for higher-skill roles as AI took over routine monitoring. This commitment built trust.

Implementation Strategy

Sensors were installed across Plant 7 while traditional inspections continued for six months. Workers logged observations as AI collected vibration, temperature, and acoustic data. When predictions aligned, confidence grew. When they didn't, both humans and machines learned.

Maria Santos became an AI trainer: "The system kept flagging Pump 3 as normal, but I knew something was wrong from the sound." Working with data scientists, she identified acoustic patterns the AI had missed. Her expertise was encoded into the algorithm, improving predictions for all shifts.

Knowledge transfer workshops replaced classroom training. Veteran workers demonstrated diagnostic techniques to junior employees and AI developers. Brown showed how he detected bearing wear through vibration patterns. These sessions preserved critical knowledge and positioned workers as partners in AI development.

New roles emerged: Predictive Maintenance Specialists combining mechanical know-how with data interpretation. These roles offered career advancement for experienced workers.

Managing Resistance

Some workers tested the system by feeding false readings. Others ignored AI alerts, trusting their senses over algorithms. Management reframed data quality as a safety and job security issue, not surveillance.

"Bad data could lead to missed failures that shut down production and threaten everyone's jobs," Rodriguez explained. The turning point came during a holiday weekend when AI predicted a critical pump failure three days before human detection would have occurred. The alert prevented a shutdown that would have cost $400,000 and forced overtime for dozens of employees.

"After that, people started listening to the machines," Santos says. "We realized the AI wasn't replacing us, it was helping us avoid the emergencies that made our lives miserable."

Critical Incident and Learning

Four months in, AI incorrectly predicted imminent failure of a conveyor motor, triggering emergency maintenance that found nothing wrong. The false alarm cost $15,000 and created skepticism.

Management responded transparently. Analysis revealed vibration readings skewed by nearby construction. Algorithms were adjusted to account for environmental noise, and findings were shared with the workforce.

"We told everyone we were all learning together; the machines, the data scientists, and the operators," Rodriguez says.

The incident strengthened trust by showing that mistakes were addressed openly and improvements followed.

Results and Impact (18 Months)

Operational Metrics

- Unplanned downtime down 42 percent
- Maintenance costs down 28 percent as repairs shifted from emergency to planned
- OEE improved from 73 percent to 89 percent
- Supply chain disruptions down 35 percent through better forecasting

Workforce Impact

- Zero involuntary job reductions
- 40 percent of maintenance staff promoted to specialist roles
- Worker satisfaction improved as emergency repairs decreased
- Knowledge transfer accelerated through AI encoding

Competitive Position

- On-time delivery performance up 25 percent
- Manufacturing costs dropped enough to win back three contracts lost to overseas competitors
- Plant 7 became GlobalParts' flagship smart factory

Why It Worked

- Framed AI as preserving and amplifying expertise
- Secured union partnership and job guarantees
- Used parallel validation to build trust
- Created career pathways through new roles
- Maintained transparent governance during errors

Questions for Reflection

- How can manufacturers position AI systems as amplifying rather than replacing worker expertise?
- What role should labor unions play in designing AI implementations that affect manufacturing workers?
- How do organizations balance the efficiency gains of AI automation with the need to maintain human skills and employment?
- When AI systems make errors in industrial settings, how

should organizations handle mistakes to maintain worker trust while improving system performance?

PUBLIC PURPOSE, DIGITAL DELIVERY: THE NOVAVISTA CASE (A HYPOTHETICAL EXAMPLE)

When government deploys AI to serve citizens, how do you balance efficiency with equity, innovation with accountability, and automation with the human touch voters expect?

The Challenge

NovaVista's City Hall faced mounting pressure in 2023. The city's 311 call center was overwhelmed; citizens waited an average of 47 minutes for answers to routine questions about garbage pickup, permits, and library hours.

City Manager Austin William was caught between competing forces: the council demanded better service without raising taxes, citizens complained about unresponsiveness, and local media criticized any tech spending as wasteful. Younger residents expected Amazon-level digital service from systems built in the 1990s.

"We needed to serve citizens better without sparking backlash about robots replacing public servants."

An AI-powered virtual assistant promised 24/7 availability and instant responses. Success required navigating union concerns, privacy regulations, political oversight, and the challenge of maintaining public trust.

The Strategic Response

William's message was clear: "NovaVista Assistant helps us serve you better, not replace the people who care about our community."

Union leaders were briefed months before launch. HR Director Lisa Park explained: "We assured them no positions would be eliminated. Staff would handle complex cases while AI managed routine inquiries."

Public engagement came next. City council meetings included citizen input sessions on privacy, bias, and human interaction. This transparency slowed the rollout but built legitimacy and trust.

Implementation Strategy

The AI assistant launched with a limited scope: about 200 FAQs covering garbage schedules, license applications, passport requirements, and library hours. Anything beyond these queries routed immediately to human staff.

"We started with boring, factual questions where AI couldn't cause controversy," says CIO Maria Santos. An AI Oversight Committee including citizens, privacy advocates, and ethics experts monitored performance. Monthly public reports detailed errors and fixes. Citizens could review their interaction logs and request human review.

"Government AI operates in a fishbowl," William noted. "Privacy and accountability aren't optional; they're mandatory."

Staff training covered technical use, public records compliance, accessibility, and equity. Employees learned to monitor AI responses for bias, ensuring equal service for all residents.

Managing Resistance

Employees feared job loss despite assurances. Citizens worried about privacy and fairness. Political opponents criticized "unproven tech" while demanding better service. Local media scrutinized every move.

NovaVista launched an AI Transparency Portal showing what systems were used, what data they processed, and how decisions were made. Every interaction disclosed: "You're chatting with NovaVista Assistant, an AI system. You can request a human at any time."

Three weeks after launch, the assistant gave incorrect business license fee information, understating costs by $200. Instead of hiding the error, the city published a correction, explained the cause, and detailed fixes.

"We used it to demonstrate our commitment to accuracy and accountability," Santos said.

Critical Incident and Learning

Two months in, a Spanish-speaking resident reported that AI responses in Spanish were less helpful than English ones. Advocacy groups raised equity concerns. Investigation confirmed the issue: the system was trained primarily on English documents.

NovaVista responded by hiring bilingual staff to expand training data, implementing automated multilingual quality checks, and establishing ongoing equity audits.

"AI fairness isn't a one-time achievement; it requires constant monitoring," Santos explained. The proactive fix, publicly reported, turned a potential crisis into proof of responsive governance.

Results and Impact (12 Months)

Service Delivery

- Average response time dropped from 47 minutes to under 2 minutes
- After-hours service requests up 340 percent
- Citizen satisfaction with responsiveness up 28 percent
- Staff availability for complex cases up 60 percent

Operational Impact

- Cost per interaction down 45 percent
- Overtime for backlog clearance down 70 percent
- Consistency of information improved across departments
- Public records processing accelerated

Political and Social Outcomes

- Public trust in city government rose modestly but steadily
- Media coverage remained largely positive
- No successful political challenges to the program
- Citizen engagement increased across demographic groups

Why It Worked

- Positioned AI as service enhancement, not job elimination
- Built union and citizen trust through transparency
- Started with low-risk, high-visibility use cases
- Established governance and equity monitoring from day one

Questions for Reflection

- How should government organizations balance AI efficiency gains with public expectations for human interaction and accountability?
- What level of transparency should public sector AI implementations provide, and how can organizations make complex algorithms understandable to citizens?
- How can government agencies ensure AI systems serve all citizens equitably while improving overall service efficiency?
- When government AI systems make mistakes affecting citizens, how should public officials respond to maintain both system credibility and political support?

TOMORROW'S WORKPLACE: THE SYNERTECH CASE (A HYPOTHETICAL EXAMPLE)

When AI becomes as ubiquitous as email, how do organizations function when every employee has a digital collaborator and traditional hierarchies dissolve into human-machine networks?

The Challenge

By 2030, SynerTech International bore little resemblance to the consulting firm that began its AI journey five years earlier. Artificial intelligence wasn't a project, it was the foundation of work. Every employee partnered with AI assistants that drafted proposals, analyzed client data, and coordinated schedules. Teams routinely included both human consultants and AI agents working seamlessly together.

"The challenge wasn't implementing AI anymore," says CEO Amanda Rivers. "It was managing an organization where the line between human and machine work had completely blurred."

Traditional management structures struggled to coordinate hybrid workflows. Approval chains couldn't keep pace with AI systems generating options in seconds. AI had become as invisible as electricity: assumed, essential, and largely taken for granted by employees who had never worked without it.

This ubiquity created new problems: AI fatigue from constant optimization prompts, heightened privacy concerns as systems required behavioral data, and the risk of human skill atrophy as employees delegated more to machines.

The Strategic Response

SynerTech launched a Skills Marketplace powered by AI to match employees with emerging opportunities, suggest personalized learning paths, and predict future capability needs. Career development shifted from ladder climbing to dynamic navigation as roles evolved quarterly with expanding AI capabilities.

Job descriptions disappeared, replaced by capability clusters: flexible combinations of human skills and AI collaboration abilities. Consultants were hired for adaptability and "AI fluency," not fixed roles.

Human work concentrated on creative problem-solving, relationship building, and ethical decision-making, while AI handled analysis, document generation, and routine communication. Employees naturally gravitated toward tasks AI couldn't replicate.

Implementation Strategy

Organizational structure flattened as AI managed coordination, resource allocation, and information flow. Middle management roles shrank dramatically; remaining leaders focused on vision, culture, and stakeholder relationships.

Decision-making moved to the edges, where AI-augmented employees had both data and authority to act. "With AI providing real-time analysis and recommendations, there was no need for decisions to flow up through layers," Rivers noted.

Teams formed dynamically based on required capabilities, often blending human consultants with specialized AI agents and coordination systems. Governance protocols defined when human judgment overrode AI, how to resolve conflicts between systems, and which decisions required human approval regardless of algorithmic confidence.

Managing the Human Element

AI fatigue emerged as employees faced constant suggestions for optimization. SynerTech introduced optimization quotas as limits on daily AI prompts and allowed individuals to adjust interaction levels.

Privacy concerns intensified as AI required behavioral data. The company deployed privacy dashboards showing what data was collected and allowing granular permission controls, reinforcing transparency and trust.

Skill maintenance became deliberate practice. "Unplugged" work sessions required consultants to solve problems without AI, preserving core capabilities and preparing for contingencies.

Critical Incident and Learning

Eight months into full integration, a system-wide algorithm failure left consultants without AI support during a critical client presentation. Several junior staff struggled with calculations and analysis they typically delegated to AI. The presentation quality suffered, exposing overdependence.

"We realized we'd created a single point of failure," Rivers reflected. SynerTech implemented analog capability requirements and regular AI-free days to maintain human proficiency. These sessions also sparked innovation as employees explored creative solutions without algorithmic shortcuts.

Results and Impact (By 2030)

Operational Excellence

- Project delivery times down 60 percent while quality scores improved
- Employee productivity per hour up 45 percent
- Innovation metrics surged as humans focused on creative work
- Client satisfaction reached record highs

Workforce Evolution

- Traditional management roles down 40 percent; coaching and mentoring roles expanded
- Skill diversity broadened as employees explored multiple competencies
- Job satisfaction improved as work became more intellectually challenging
- Continuous learning became standard expectation

Organizational Capabilities

- Decision-making speed accelerated without sacrificing quality
- Cross-functional collaboration improved through AI-mediated coordination
- Knowledge sharing scaled as AI captured and distributed insights
- Competitive advantage shifted from having information to interpreting it wisely

Why It Worked

- Built a culture of continuous learning and adaptability
- Redefined roles around human-AI collaboration
- Flattened hierarchy to match AI-driven speed
- Maintained resilience through privacy safeguards and skill retention

Questions for Reflection

- How should organizations maintain human skills when AI handles most routine cognitive work?
- What governance structures work best for hybrid human-AI teams where decision-making authority is distributed?

- How can companies balance AI personalization benefits with employee privacy concerns as systems become more integrated?
- When AI becomes organizational infrastructure like electricity, how do leaders maintain human agency and avoid over-dependence on algorithmic systems?

WHAT HIGH-PERFORMING AI ORGANIZATIONS DO DIFFERENTLY

Across industries, the companies that turn AI from promise into performance share a common playbook. Their advantage is how they lead, organize, and learn.

In organizations that keep extracting value from AI, you can usually trace choices back to a sentence everyone knows by heart. People use it to explain why a feature was cut, why a pilot continued despite early stumbles, why a model was retired while it still worked. The sentence does not sound like a slogan. It reads like a promise to customers, a standard for employees, and a boundary the firm will not cross.

Watch the senior team for a week and the posture becomes clear. They do not outsource learning. They show their work in public, ask naïve questions, and let their calendars betray the time they spend with users and with the people closest to the process. When uncertainty rises, they widen the circle rather than tightening it, and they treat momentum as something built with updates, not pronouncements.

The culture feels practical rather than heroic. Small experiments are easy to start and easier to stop. People are expected to surface doubt early, especially when an AI output seems plausible. Meetings end with one or two things to try, one thing to measure, and a note on what will be abandoned if the data disagrees. The lesson that sticks is not that failure is welcome, but that learning must be fast and traceable.

Teams that adopt AI well tend to design the human parts with the same care as the model. Stakeholders see early versions before the paint is dry. Training is hands on and local to the workflow. Feedback loops are short, and what returns from the field actually changes the next release. Adoption looks less like change management and more like craft, taught through repetition and coaching.

Governance shows up as clarity rather than ceremony. People know which risks are never acceptable, what evidence unlocks a wider rollout, who has authority to pause a system, and how to appeal a decision. The effect is speed, not drag, because ambiguity has a higher cost than a lightweight checklist.

The scorecard is wider than finance. Leaders keep an eye on trust, skills, safety, and customer outcomes, and they let those signals move investment. When something improves the P&L while eroding capability or confidence, it is treated as a temporary win with long term risk, not a triumph.

The cadence is continuous. Models, data pipelines, and processes are refreshed on a schedule the company can feel. Wins are retired when they plateau. The next wave is in motion before the last one peaks, so change reads as the normal state of the system rather than a series of campaigns.

If there is a pattern, it is that technology is asked to perform at the same level as management. The tools are strong, but the habits are stronger.

SUMMARY

The journey through these industry scenarios reveals a fundamental truth: AI transformation success patterns transcend sectors, technologies, and organizational structures. Whether in banking's regulated corridors or manufacturing's shop floors, healthcare's clinical settings or government's public forums, the same principles determine which organizations thrive and which merely survive.

Chapter takeaways:

- AI transformation is underway in every industry, but most organizations struggle to achieve large-scale impact because the main barriers are people, process, and culture.
- Success depends on clear executive vision, early and ongoing employee involvement, ethical foresight, and a culture that supports experimentation and learning.
- Effective change management starts with small, low-risk pilots, builds trust through transparency and inclusion, and focuses on designing for human-AI collaboration rather than just automation.
- Organizations that thrive with AI institutionalize learning from mistakes, treat resistance as valuable feedback, and continuously adapt.
- Across all sectors, leadership, engagement, ethics, and adaptability are the key determinants of sustainable AI success.

Chapter 7 takes these real-world lessons and distills them into actionable frameworks. Where Chapter 6 showed what success looks like across industries, Chapter 7 provides the blueprint for achieving it. We move from observation to application, from understanding patterns to implementing practices.

The integration framework ahead synthesizes insights from financial services' governance models, healthcare's stakeholder engagement approaches, manufacturing's workforce evolution strategies, and government's transparency mechanisms. It transforms scattered best practices into a coherent methodology any organization can adapt to its unique context.

Most importantly, Chapter 7 addresses the question every leader asks after seeing these examples: "How do I actually do this in my organization?" The answer lies not in copying any single industry's approach, but in understanding the universal principles that make transformation possible and applying them with wisdom and courage to your specific situation.

7

PRACTICAL FRAMEWORKS FOR SUCCESSFUL AI-DRIVEN CHANGE

A January 2025 Kearney study found that while 89 percent of CEOs acknowledge the strategic importance of AI, only one in four feel fully prepared to integrate it. This stark disconnect reveals a truth leaders must confront: implementing AI is primarily an organizational transformation that demands systematic change management.

Only 14 percent of companies worldwide report complete confidence in their AI projects. The cost of delay compounds daily as AI capabilities accelerate exponentially. Organizations that hesitate risk falling irreversibly behind competitors who master AI integration.

This chapter provides C-suite executives, transformation leaders, and change managers with practical frameworks for bridging the AI implementation gap. We blend proven organizational change management principles with AI-specific insights from recent transformations to equip you with actionable tools for guiding their organizations through AI-driven change in a structured, ethical, and sustainable way.

The frameworks ahead address seven critical dimensions of AI transformation. First, we examine organizational readiness assessment, identifying strengths, gaps, and the urgency for change. Next, we explore comprehensive AI strategy development and roadmap creation, ensuring initiatives align with business goals through phased implementation. Leadership and communication plans follow, emphasizing executive sponsorship and transparent dialogue as trust-building cornerstones.

We then detail workforce training and upskilling programs that transform apprehension into empowerment. Ethical governance frameworks establish guardrails keeping AI initiatives responsible and trustworthy. We address resistance management and employee engagement, showing how to convert skepticism into support using models like ADKAR and Kotter's principles.

Finally, we focus on sustaining momentum through continuous improvement and success measurement. AI adoption isn't a one-time project but an ongoing evolution requiring metrics, feedback loops, and cultural reinforcement.

Each framework builds on established change methodologies such as Kotter's 8-Step Process, Prosci's ADKAR model, and McKinsey's 7S framework. Real-world examples ground every concept in practical application.

The journey begins with honest assessment of where you stand today and clarity about where AI can take you tomorrow. In an age of AI dominance, companies must humanize the swarm: aligning people, processes, and values with powerful new technologies. Without structured change management, even the most sophisticated AI initiatives will fail. With the right frameworks, organizations can navigate complexity, build capability, and capture AI's transformative potential.

AI transformation demands bold vision coupled with systematic execution. The pages ahead provide the roadmap.

GAUGE YOUR ORGANIZATION'S AI READINESS

Success requires confronting uncomfortable truths about your organization's AI preparedness.

Before rushing into implementations, leaders must strip away wishful thinking and assess foundational readiness across strategy, culture, skills, and infrastructure. An honest evaluation prevents costly failures while identifying critical gaps that demand attention before diving into AI deployment.

The McKinsey 7S framework provides a comprehensive lens for organizational assessment. Each element reveals potential barriers to AI success:

Strategy: Does AI explicitly support business objectives? Organizations must verify that AI initiatives align with strategic goals rather than pursuing technology for its own sake. AI strategy should complement core business priorities, not exist as a standalone experiment.

Structure: Can your organization support cross-functional AI teams? Traditional hierarchies often impede the agile collaboration AI requires. Business units must be prepared to work seamlessly with data science teams, breaking down silos that fragment AI efforts.

Systems: Legacy processes and infrastructure frequently block AI integration. Organizations need modern data pipelines, cloud platforms, and cybersecurity measures capable of supporting AI at scale. Critical questions include: Do you have necessary data architectures? Can current systems integrate AI solutions effectively?

Shared Values: Organizational culture shapes AI destiny. Companies that value innovation, data-driven decisions, and experimentation adapt more readily to AI. Resistance often stems from cultures that punish failure or resist change, a strong indicator that cultural transformation must precede technical implementation.

Style: Leadership approach determines AI success. Inclusive, forward-thinking leadership styles build trust and adaptability crucial for AI adoption. Command-and-control management clashes with AI's iterative, experimental nature.

Staff: Beyond technical specialists, organizations need broad AI literacy. Marketing teams must understand AI segmentation tools. Managers need skills for overseeing AI-augmented processes. Finance professionals require competence with AI forecasting models.

Skills: The workforce needs both technical and adaptive capabilities. While data science expertise matters, equally critical are skills like critical thinking, problem-solving, and creativity - human abilities that complement AI's computational power.

When assessing your organization's AI readiness, consider the following factors:

Start with your data: Ask yourself: Is your data truly ready for AI, or are you still wrestling with silos and inconsistent definitions? The organizations that win with AI treat data as a strategic asset. They make it accessible across teams, enforce quality controls, and ensure governance is more than a policy on paper.

Examine your technology backbone: Look beyond what's working today. Do you have the cloud resources, hardware, and scalable platforms to support AI at scale? Can your legacy systems integrate with new solutions, or are they holding you back? The leaders who get ahead invest in infrastructure with tomorrow's needs in mind, not just today's.

Lead from the front: Your commitment sets the tone. Are you articulating a clear vision for AI that ties directly to business goals? Are you allocating real budgets for experimentation, and making AI a boardroom priority? When you show urgency and clarity, your organization follows.

Shape a culture that's ready to adapt: Technology alone won't move your organization forward. Take a hard look at how your teams feel about AI. Are they curious, or cautious? Do you have internal champions who can help others learn? Encourage experimentation, reward learning, and make it clear that AI is here to augment, not replace. Address fears openly and nurture enthusiasm.

Make change management a core strength: If your teams are already comfortable with agile methods and iterative improvement, you're ahead of the curve. But don't stop there. Dedicate resources to change management, draw on your history of successful transformations, and build flexibility into every process. For high-performing organizations, change is a habit.

A thorough AI readiness assessment requires a structured approach that goes beyond technology to address the full spectrum of organizational factors. The 5P framework offers a holistic lens for evaluating your organization's preparedness, ensuring that every critical dimension is considered before embarking on AI transformation:

Purpose: Why pursue AI? Clear strategic rationale prevents scattered efforts.

People: Who drives change? Map stakeholder readiness, identify champions and skeptics, understand skill gaps.

Process: How will work flow change? Document current processes and envision AI integration points.

Platform: What technology exists? Assess tools and infrastructure against AI requirements.

Performance: How will you measure success? Define metrics before implementation begins.

Leaders who succeed with AI transformation deploy a suite of practical assessment tools to surface strengths and gaps before launching major initiatives. AI readiness scorecards provide a structured way to rate each dimension of preparedness, while stakeholder surveys reveal the true sentiment (optimism, skepticism, or anxiety) rippling through the organization. Skills inventories map current capabilities, highlighting both hidden talent and critical gaps. Technology audits go beyond surface-level checks, probing whether infrastructure can support the demands of modern AI. Meanwhile, process documentation uncovers where automation could deliver the greatest impact, often revealing inefficiencies that have long gone unnoticed.

Yet even the most sophisticated tools are only as valuable as the candor with which leaders interpret their results. Certain warning signs consistently signal insufficient readiness. An AI strategy that drifts away from core business priorities is a common pitfall, as is fragmented data trapped in incompatible systems. Limited executive understanding or lukewarm support can stall momentum before it begins. Employee fear often festers in the absence of clear communication. The lack of an ethical governance framework leaves organizations exposed to reputational and regulatory risk. And when change management is treated as an afterthought, even the best technology investments struggle to take root.

To ensure a truly robust AI readiness assessment, it's essential to look beyond internal factors and consider the broader landscape in which your organization operates. The Technology-Organization-Environment (TOE) framework brings this external perspective into focus by examining three critical dimensions:

- Organizational context (structure, processes, culture)
- Technological context (availability, complexity, compatibility)
- Environmental context (competitive pressure, regulatory requirements)

As pilots launch and capabilities mature, regular reassessment guides course corrections. Organizations typically discover they're stronger in some areas while weaker in others. Create action plans addressing identified gaps. Low data quality? Launch governance initiatives.

Cultural resistance? Begin awareness campaigns. Skills shortage? Design training programs. Address weaknesses systematically while leveraging strengths.

Most organizations rate between levels 2-3 on AI maturity scales, indicating basic capabilities but limited strategic integration. This baseline helps set realistic expectations and development priorities.

Remember: Perfect readiness isn't the goal. Understanding current state, acknowledging gaps, and creating plans to address them matters more. The assessment provides a roadmap for building AI capability systematically rather than haphazardly.

Successful AI transformation starts with honest self-evaluation. Organizations that assess thoroughly, address gaps methodically, and build on existing strengths position themselves for sustainable AI success. Those that skip assessment or ignore findings typically face expensive failures that could have been prevented.

The journey to AI maturity begins with knowing where you stand today. Only then can you chart a realistic path to where AI can take your organization tomorrow.

LEAD THE AI TRANSFORMATION

Even the most brilliant AI strategy stalls without leaders who can navigate organizational politics, address employee concerns, and maintain momentum through inevitable setbacks.

Leadership is the decisive factor in whether AI transformation delivers lasting value or fizzles out as an expensive experiment. Consider Microsoft's journey under Satya Nadella. When Nadella declared that AI would reshape every product and process, he reorganized the company, restructured incentives, and became the visible champion of change. This level of executive commitment is what distinguishes organizations that realize AI's promise from those that simply dabble.

Executive sponsorship, in practice, is never just ceremonial. The leaders who drive real change allocate scarce resources (whether budget, talent, or time) to the initiatives that matter most. They remove organizational barriers, making it possible for teams to move quickly and decisively. When projects stall or struggle, these leaders make the hard calls, whether that means pivoting, doubling down, or pulling the plug. They communicate the strategic importance of AI consistently, reinforcing its relevance at every level of the organization. They hold teams accountable for measurable outcomes, and perhaps most importantly, they model the behaviors they expect from others, adopting new tools, learning in public, and demonstrating adaptability.

The structure of leadership itself shapes the trajectory of transformation. Three models have emerged in practice, each with distinct strengths and trade-offs. In the centralized approach, a Chief AI Officer or Chief Digital Officer owns the enterprise-wide AI strategy. This model offers clear accountability and a unified direction, making it well-suited for organizations at the beginning of their AI journey. However, it can risk disconnecting from the nuanced needs of individual business units.

A federated model distributes leadership, allowing business units to lead their own AI initiatives while a central team provides coordination and guidance. This approach keeps AI close to business needs and accelerates adoption, but it can lead to fragmentation or duplicated efforts if not carefully managed. It tends to work best in organizations with a mature understanding of AI and a track record of successful deployments.

The hybrid model seeks to balance consistency with agility. Here, a central strategy team sets direction and standards, while embedded execution teams within business units drive implementation. This structure is ideal for large, complex enterprises with diverse needs, but it requires sophisticated governance to manage the inherent complexity and potential for internal conflict.

Translating leadership intent into organizational action requires structure. Effective governance mechanisms provide that structure, turning aspiration into execution.

The **AI Steering Committee** brings together C-suite executives, business leaders, and technical experts. They meet monthly to make strategic decisions, allocate resources, and remove barriers. This group ensures the entire organization moves in concert.

The **AI Center of Excellence** serves as the engine room. Data scientists, engineers, business analysts, and ethicists form a central team that builds capabilities, shares best practices, and supports projects across the enterprise. In its early stages, the center leads projects directly. Over time, it shifts toward enabling others, equipping teams with the tools and guidance to succeed independently.

The **AI Ethics Board** plays a critical role in safeguarding trust. Composed of internal experts, external advisors, and legal counsel, the board develops policies, reviews high-risk use cases, and investigates incidents. Its authority must be real. Without the power to veto problematic applications, ethical oversight becomes symbolic rather than substantive.

Governance alone isn't enough. Communication is what builds momentum and earns trust.

Frame the narrative positively, positioning AI as a source of opportunity rather than a threat. Messages like "AI augments human capability" and "We're becoming AI-first to better serve customers" shape perceptions and inspire confidence.

Address concerns directly, acknowledging legitimate fears while offering concrete reassurance. Job security worries call for visible commitments to retraining. Privacy concerns demand clear safeguards. Technical complexity requires education that is not just accurate but accessible.

Celebrate progress publicly, knowing that recognition fuels momentum. Pilot successes, customer impact stories, and team achievements deserve the spotlight. Even failures, when shared openly, become learning moments that strengthen resilience.

Use multiple channels to ensure the message lands. Town halls align the vision. Team meetings allow for deeper discussion. Email updates track progress. Video messages create emotional connection. Workshops offer hands-on engagement. No single channel suffices; effective communication is layered and persistent.

Success often hinges on those who sit between strategy and execution. Middle managers, frequently overlooked in AI transformations, are the ones who translate vision into daily operations. They influence morale, shape workflows, and control how resources are deployed. When they're left out, even the most ambitious initiatives can stall.

Bringing them into the fold early changes the dynamic. When they understand the strategy from the outset, they're better equipped to guide their teams through change. Their roles evolve, and with that evolution comes the need for new skills, especially in managing AI-driven projects. Support systems matter. Metrics that recognize their contribution to transformation, and spaces where they can voice concerns, help sustain their engagement.

In parallel, change champions emerge. These are the people others listen to, not because of their title, but because of their credibility. They're curious about technology, comfortable with ambiguity, and able to explain complex ideas in ways that resonate. They don't give up easily. Their persistence, especially when things get messy, helps build grassroots momentum.

Resistance is part of the process. It often comes from fear: fear of losing a job, fear of being judged by algorithms, fear of not understanding the tools. Sometimes it's about power, and the discomfort of shifting dynamics. Leaders who navigate this terrain with empathy make a difference. Listening helps. So does offering evidence that speaks to specific concerns, and creating moments where skeptics can see AI's value for themselves.

Culture is the foundation. Without it, technology struggles to take root. The shift is subtle but essential. Decisions move from gut instinct to data-informed judgment. Teams start collaborating across functions, and technology becomes a source of value rather than a cost center. The narrative shifts from humans versus machines to humans working with machines.

Leadership presence matters throughout. Nail the kickoff, but have regular touchpoints (quarterly updates, AI-focused business reviews, site visits, customer meetings, board presentations) signal commitment. They keep the energy up and the direction clear.

The most effective leaders make AI transformation personal. They use AI tools themselves, share their learning journey, and model continuous adaptation. When employees see leaders embracing change, resistance decreases and adoption accelerates. Strong leadership sets the vision, but execution depends on employees who possess the skills and confidence to work alongside AI systems. Systematic capability building becomes essential for AI success.

EQUIP YOUR WORKFORCE FOR THE AI ERA

The AI skills gap threatens more organizations than budget constraints.

While companies rush to implement AI technologies, human capacity lags dangerously behind. A global survey revealed 87 percent of companies face AI skill shortages, yet only 23 percent have implemented comprehensive training programs. This disconnect creates a ticking time bomb for digital transformation efforts.

AI literacy isn't about creating armies of data scientists. Success requires appropriate skill levels across all organizational roles from executives making strategic decisions to frontline workers operating AI-enhanced tools. The goal is to build an AI-fluent organization where everyone understands their role in the AI-augmented workplace.

Skill requirements shift as you move through the layers of an organization. What executives need to lead transformation differs from what frontline employees need to make it work.

At the top, strategic clarity is essential. Executives must be able to shape and evaluate AI strategy, weigh ethical implications, and prioritize investments. Their decisions hinge on frameworks for assessing risk and guiding transformation. Leadership here is more about steering the ship with confidence and foresight.

Managers operate closer to the action. Their role demands oversight of AI projects and the ability to manage hybrid teams that blend human and machine capabilities. Change management becomes a daily task. They must ensure ethical implementation, track return on investment, and adjust course when needed.

Professionals need to understand how AI applies to their specific field, interpret data effectively, and use AI tools with fluency. Their work often involves redesigning processes and ensuring the quality of AI outputs, tasks that require both technical skill and contextual judgment.

Technical teams build the infrastructure. Their work includes developing machine learning models, engineering data pipelines, integrating systems, and optimizing algorithms. They manage the platforms that make AI possible, often working behind the scenes but with outsized impact.

Frontline employees interact with AI directly. They operate tools, handle exceptions, and provide feedback that helps refine systems. Their understanding of process compliance and their ability to explain AI-driven decisions to customers are critical to trust and adoption.

Learning pathways work best when they reflect the diversity of starting points. Not everyone enters the AI conversation with the same background, and effective programs account for that.

At the broadest level, AI awareness is universal. Everyone in the organization benefits from a foundational understanding of what AI is, how it fits into the company's strategy, and what ethical considerations come into play. When people grasp the personal and professional impact of AI, they're less likely to resist change out of fear.

For those using AI tools in their day-to-day work, training must be practical. AI application focuses on the tools themselves, how they integrate into workflows, and how to interpret outputs. It also covers how to recognize errors and escalate issues appropriately. The goal is to build confidence in real-world use.

Leaders need a different set of capabilities. AI management involves evaluating projects, building teams that can work effectively in hybrid environments, and measuring performance in new ways. It also means understanding the risks unique to AI and communicating clearly about initiatives that may be unfamiliar or complex.

Specialists go deeper. AI development requires advanced technical skills and platform fluency. These individuals stay current with emerging capabilities and best practices, and they know how to translate research into tools that solve real problems. Their work bridges innovation and execution.

Learning works best when the method matches the goal. Different objectives call for different **delivery formats**, and the most effective programs blend them thoughtfully.

Scalable foundation-building often begins with self-paced digital learning. Microlearning modules, interactive simulations, and progress tracking offer flexibility and reach. Certificate programs add structure, while quizzes reinforce retention. For building broad awareness, this approach is both efficient and cost-effective.

When deeper capability is the aim, instructor-led workshops provide the necessary depth. Hands-on exercises, group problem-solving, and expert facilitation create an environment where complex skills can take root. Peer learning and immediate feedback accelerate growth and build confidence.

For more personalized development, mentoring programs offer something different. One-on-one guidance, career planning, and project-based learning help individuals navigate their unique paths. These relationships transfer tacit knowledge and build networks, critical ingredients for developing future AI leaders.

Some lessons require experimentation. Innovation labs create space for trial and error. Sandbox environments, proof-of-concept development, and cross-functional collaboration foster comfort with uncertainty. Rapid iteration and tolerance for failure help teams learn by doing.

And when internal resources reach their limits, external partnerships fill the gaps. Collaborations with universities, vendor certifications, conference participation, and consultant engagements bring in specialized expertise. These connections extend the organization's learning capacity and keep it aligned with the frontier of AI practice.

Curriculum design is more than content. The most effective learning programs are built on principles that reflect how adults learn, work, and change.

It starts with relevance. Learners engage when the material connects to their actual work. Using company data, solving real problems, and featuring familiar processes makes the experience tangible. When current tools are part of the training and the benefits are clear, participation becomes purposeful. Adults don't learn to apply.

Progression matters just as much. Complexity should build gradually. Fundamentals come first, followed by deeper concepts and specialized paths. Frequent assessments and practice opportunities help learners stay on track. Overloading them too early leads to disengagement. A well-paced journey keeps them moving forward.

To sustain attention, engagement must be intentional. Interactive exercises, compelling case studies, and visual demonstrations bring ideas to life. Gamification adds energy, while social learning creates connection. Passive content, no matter how well-written, rarely sticks.

Learning must also lead to doing. Application is where theory becomes value. Practical projects, integration into daily work, access to tools, and performance feedback all reinforce the learning. Without these, even the best-designed curriculum risks becoming shelfware.

Yet even well-funded programs falter. Common pitfalls include generic content that ignores role-specific needs, an overemphasis on technical depth at the expense of business relevance, and too few opportunities for hands-on practice. Some programs lack post-training support or fail to align with the tools people actually use. Others underestimate the importance of change management, assuming knowledge alone will drive adoption.

Measurement helps close the loop. Tracking both activity and impact ensures that learning translates into outcomes.

Activity metrics include enrollment and completion rates, assessment scores, skill demonstrations, satisfaction ratings, and support usage. These show who is participating and how they're progressing.

Impact metrics go further. They track AI tool adoption, process improvements, error reductions, innovation submissions, and productivity gains. These indicators reveal whether learning is driving transformation.

Career pathways are shifting. The traditional ladder that was built on static skill sets and predictable roles is giving way to something more dynamic. AI reshapes what roles mean and how they evolve.

Business analysts begin to take on product ownership, guiding AI solutions from concept to deployment. Operations managers reimagine workflows as automation architects. Customer service representatives become experience designers, crafting interactions that blend human empathy with machine efficiency. Financial analysts move toward algorithmic trading, and HR generalists grow into people analytics experts. These transitions are already underway.

One financial services firm offers a glimpse into what comprehensive execution looks like. Their "AI for All" program spans the organization. Executives participate in immersion sessions focused on strategic implications. Managers engage in simulations that mirror real-world leadership challenges. Professionals follow certification tracks tailored to their functions. Technical specialists attend bootcamps to deepen platform expertise. Frontline employees receive training on the specific tools they use daily. It's role-aware, layered, and continuous.

That continuity matters. Learning is a rhythm. Annual skill assessments surface emerging gaps. Quarterly curriculum updates keep pace with new capabilities. Monthly lunch-and-learns share innovations across teams. Weekly tips maintain momentum. Over time, this cadence builds habits.

In this environment, curiosity becomes currency. Experimentation is encouraged. A growth mindset is a competitive necessity. Organizations that build systematic learning capabilities create advantage in landscapes shaped by rapid technological change.

Absorbing, distributing, and acting on new knowledge becomes the new differentiator. It's no longer enough to track learning activities. What matters is learning agility: how fast a company learns and how effectively it moves, interprets, and recombines knowledge.

MEASURING KNOWLEDGE FLOW AND ADAPTIVE CAPACITY

Organizations are beginning to rethink how they measure learning. Traditional metrics focus on activity: who attended, what was completed, how many modules were launched. But in environments shaped by AI, activity alone is no longer enough. The real question is how quickly an organization can absorb new knowledge, distribute it across teams, and apply it to decisions.

New indicators are emerging. **Time-to-absorb insight** measures how long it takes for new information or AI-generated output to be interpreted, acted upon, and integrated into decision-making. This metric reflects not just access to data but the speed of understanding.

Knowledge flow index tracks how effectively insight moves across departments, levels, and systems. It can be assessed through sentiment analysis, social graph mapping, or audits of communication patterns. When knowledge flows freely, organizations adapt faster.

Recombination velocity captures how quickly diverse knowledge elements (i.e., past projects, AI outputs, external research) are merged to create novel solutions or reused in new contexts. It reflects the organization's ability to innovate by repurposing what it already knows.

These metrics shift the focus from individual expertise to collective learning speed. They also expand the role of organizational change management. Instead of focusing solely on adoption, OCM becomes a steward of insight movement.

To embed these metrics into practice, organizations can build AI dashboards that highlight where knowledge flows are stalling. They can monitor the reuse of internal prompts, playbooks, and insights across functions. They can track idea lineage to see how quickly small ideas evolve through feedback and iteration into scaled value.

Change professionals should treat these as leading indicators of resilience, speed, and innovation capacity in an AI-saturated world. As organizations accelerate their capacity to learn and adapt, they must also ensure that this speed doesn't outpace their ethical judgment. A high-velocity knowledge ecosystem without responsible guardrails can amplify risks as quickly as it scales innovation.

BUILD GUARDRAILS FOR RESPONSIBLE AI

Ethics is a competitive advantage. When Amazon scrapped its AI recruiting tool for showing gender bias, when Facebook faced scrutiny over discriminatory ad targeting, when criminal justice algorithms perpetuated racial disparities, the message became clear: good intentions without systematic governance leads to reputational damage, legal liability, and lost trust.

Responsible AI requires operationalizing principles that most organizations claim to value: fairness, transparency, accountability, privacy, security. The gap between stating these principles and implementing them effectively determines whether AI becomes an asset or a liability.

Core AI principles require precise definitions. Without clarity, organizations risk inconsistent implementation and diluted accountability.

Fairness means AI systems must avoid discrimination. This involves testing for bias across demographic dimensions, monitoring outcomes for disparate impact, and correcting imbalances when they appear. Fairness is not a static goal, it requires ongoing vigilance.

Transparency ensures that stakeholders understand how AI systems influence decisions. This includes explaining model logic in accessible terms, documenting data sources and limitations, and disclosing when AI plays a role in shaping outcomes. Transparency builds trust and enables informed oversight.

Accountability assigns ownership. Each AI system must have a clearly identified individual or team responsible for its performance. That ownership includes the ability to explain decisions and the authority to intervene when necessary. Without accountability, responsibility becomes diffuse and risk increases.

Privacy protects personal data. Effective safeguards include obtaining informed consent, minimizing data collection, securing information appropriately, and deleting data when it is no longer needed. Privacy is not just a legal requirement but a foundation for ethical AI use.

Security defends AI systems from manipulation. This involves protecting the integrity of training data, preventing adversarial attacks, monitoring for unusual patterns, and maintaining strong access controls. Security is essential to maintaining system reliability and public confidence.

Governance structures play a critical role in translating AI principles into operational practice. Without them, even well-defined values risk becoming aspirational rather than actionable.

The **AI Ethics Board** serves as a central anchor. Composed of ethicists, legal experts, technologists, business leaders, and external advisors, the board develops policies, reviews high-risk projects, investigates incidents, educates employees, and communicates publicly about AI practices. Its diversity ensures that decisions reflect multiple perspectives and remain grounded in both technical and societal realities.

A **risk assessment framework** provides structure for evaluating AI projects. It considers impact severity, probability factors, and mitigation options. Applications with significant consequences (such as healthcare diagnosis, credit decisions, or employment screening) require enhanced scrutiny. The framework ensures that risk is not just acknowledged but actively managed.

Development checkpoints embed ethical review into the lifecycle. These mandatory reviews occur at key stages, including design assessment, pre-deployment testing, and post-launch monitoring. By placing gates throughout the process, organizations prevent ethical considerations from becoming afterthoughts. Instead, they become part of how systems are built.

An **incident response process** prepares the organization for the unexpected. Clear procedures guide escalation, investigation, stakeholder communication, and remediation. When AI systems behave unpredictably, the response must be swift, transparent, and accountable.

Together, these structures move organizations from principle to practice. They create the conditions for responsible AI.

Policy frameworks guide daily decisions by translating abstract principles into operational clarity. Without them, ethical intent can falter in the face of routine complexity.

An **acceptable use policy** defines what AI can and cannot do within the organization. Some companies explicitly prohibit using AI for performance terminations or surveillance without human oversight. These boundaries help prevent misuse and reinforce trust.

Data governance standards specify how data used in AI systems should be collected, stored, used, and shared. They include requirements for data quality, consent management, retention periods, and deletion procedures. Strong governance ensures that data practices remain aligned with legal and ethical expectations.

Algorithmic audit requirements mandate regular testing of AI systems for bias, accuracy, and fairness. These reviews may be conducted internally or by independent third parties, especially for high-stakes applications. Audits help maintain accountability and surface unintended consequences before they scale.

Human-in-the-loop guidelines identify scenarios where human oversight is essential. Decisions that affect people's lives such as medical diagnoses, loan approvals, or hiring outcomes often require human review. These safeguards ensure that automation does not replace judgment where it matters most.

Together, these policies create a foundation for responsible AI. They offer clarity, reinforce accountability, and help organizations navigate the complexity of daily decision-making with confidence.

Regulatory compliance grows increasingly complex. The EU's AI Act categorizes AI uses by risk level, imposing different requirements for each category. GDPR affects AI using personal data. Sector-specific regulations add additional layers. Organizations must:

- Track evolving regulations globally.
- Assess compliance gaps systematically.
- Implement required controls proactively.
- Document compliance efforts thoroughly.
- Prepare for regulatory audits.

Technical implementation of ethics requires specific tools and practices:

Bias Testing Toolkits: Software libraries that measure model fairness across different demographic groups, identify sources of bias, and suggest mitigation strategies.

Explainability Methods: Techniques like SHAP values or LIME that help interpret black-box models, generating human-understandable explanations for individual predictions.

Privacy-Preserving Techniques: Methods such as differential privacy, federated learning, or homomorphic encryption that enable AI development while protecting individual privacy.

Audit Trails: Comprehensive logging of AI decisions, data lineage tracking, and model version control to enable accountability and investigation when issues arise.

Effective AI governance depends on meaningful stakeholder engagement. Without it, policies risk being disconnected from the people and communities they affect. Engagement is not a one-time consultation, it must be built into the governance process as a continuous feedback loop.

Employees offer critical insight. Regular surveys and open feedback channels surface concerns from those working with AI systems every day. In some organizations, employee-led ethics committees have identified risks that leadership might otherwise overlook. Their proximity to operational realities makes their input especially valuable.

Customers expect transparency. Clear communication about how AI is used in interactions builds trust. Providing opt-out options and mechanisms to appeal AI-driven decisions reinforces accountability and gives users a sense of control. Transparency also helps customers understand the value AI adds to their experience.

External advisory boards bring independent oversight. These groups (often composed of ethicists, technologists, legal scholars, and industry experts) review policies, assess practices, and offer unbiased perspectives on emerging ethical challenges. Their independence helps ensure that governance remains credible and balanced.

Community consultation becomes essential when AI systems impact broader populations. Proactive engagement with affected groups helps identify potential harms before deployment and ensures governance reflects lived experience. These conversations often reveal unintended consequences that internal teams may not anticipate.

Several governance failures are common and avoidable. Treating ethics as a compliance checkbox undermines credibility. Policies without enforcement mechanisms lack impact. Overreliance on technical fixes ignores the human dimensions of ethical risk. Excluding stakeholder voices weakens legitimacy. Underestimating the resources required for oversight leads to gaps in execution. These missteps can erode trust and stall adoption.

Ethical AI is not about perfection. It is about managing risk systematically and improving continuously. Organizations that build robust governance frameworks earn stakeholder trust, reduce legal exposure, and often discover that constraints can drive innovation rather than limit it. Ethical rigor becomes a source of strategic advantage.

The future belongs to organizations that embed ethics into AI development from the start. Retrofitting after problems emerge is costly and ineffective. Responsible AI is about building systems that reflect organizational values and contribute to human flourishing. When ethics is intrinsic, innovation becomes more inclusive, resilient, and sustainable.

With guardrails in place to ensure responsible AI deployment, organizations face perhaps their greatest challenge: the human response to technological change. Even the most ethically sound and well-designed AI systems fail when people refuse to adopt them. The next critical step transforms resistance from a barrier into a catalyst for better implementation.

TRANSFORM RESISTANCE INTO MOMENTUM

Resistance signals organizational health, not dysfunction. When employees push back against AI initiatives, they're often highlighting legitimate concerns that, if ignored, could derail entire transformations. A pharmaceutical company discovered this when lab scientists resisted an AI drug discovery platform because the system couldn't explain its molecular predictions, threatening regulatory compliance.

Understanding the **sources of resistance** is essential for designing targeted interventions. When organizations address concerns directly, they not only reduce friction but often unlock unexpected enthusiasm.

Fear of Job Displacement remains the most visceral concern. "Will AI replace me?" is a question that echoes across industries. One bank responded by showing mortgage processors how AI handled routine applications, freeing them to focus on complex cases requiring human judgment. As processing speed increased, business grew, and employment rose with it.

Loss of Professional Identity can be equally destabilizing. Experts may fear that their hard-earned knowledge will be sidelined. Senior traders initially resisted algorithmic systems until they were repositioned as "AI strategy architects," responsible for teaching algorithms market nuances. Their expertise didn't diminish, it evolved.

Competence Anxiety often stems from unfamiliarity with technical systems. Insurance adjusters pushed back against photo-based AI damage assessments until hands-on training demystified the technology. Once they saw AI as a tool rather than a threat, resistance turned into advocacy.

Autonomy Erosion triggers discomfort when professionals feel their judgment is being replaced. Physicians rejected clinical decision support tools until interfaces were redesigned to present AI insights as supplemental information. Preserving decision-making authority proved essential to adoption.

Workload Concerns arise when AI unintentionally increases complexity. Customer service representatives found that chatbots escalated only the most difficult cases, increasing their burden. A redesign of escalation protocols, paired with targeted training, helped restore balance and reduce frustration.

Ethical Objections reflect principled resistance. HR professionals refused to use AI screening tools until they were shown bias testing results and given control over algorithm parameters. Transparency and agency helped rebuild trust and open the door to responsible use.

The ADKAR framework offers a structured approach to change management, especially in AI adoption. It helps organizations move from awareness to sustained behavior change by addressing the human side of transformation.

Awareness begins with understanding why change is necessary. Thirty-one percent of employees admit to sabotaging AI efforts through non-use. Communicating the rationale (competitive pressure, efficiency demands, strategic imperatives) must be done in terms that resonate personally. Without relevance, awareness remains superficial.

Desire builds on awareness but goes deeper. People must want the change, not just understand it. Addressing "What's in it for me?" questions is essential. Connecting AI adoption to personal benefits like less repetitive work, new skill development, and career growth helps shift mindset from resistance to engagement.

Knowledge equips people with the tools to act. Understanding how to change requires comprehensive training, accessible resources, and opportunities to practice. Expert support reinforces learning and builds confidence. Without knowledge, even motivated employees struggle to contribute.

Ability turns knowledge into execution. Knowing what to do is not the same as being able to do it. Employees need the right tools, clear processes, realistic expectations, and time to build proficiency. Removing environmental barriers (technical, procedural, or cultural) is critical to success.

Reinforcement ensures change sticks. Without it, people revert to old habits. Recognition for adoption, usage metrics, adjustment support, and celebration of success all help sustain momentum. Reinforcement must be ongoing, not a one-time event.

Engagement strategies help convert skeptics into champions. When resistance is met with inclusion, transparency, and support, it often transforms into advocacy.

Co-creation Approach invites skeptics into the design process. A manufacturing company faced opposition from quality inspectors over automated defect detection. By involving them in training the AI (teaching it what defects looked like) their resistance shifted to ownership. Participation gave them a stake in the outcome.

Pilot Program Inclusion leverages credibility. Selecting respected skeptics for early implementations can turn critics into advocates. A retail chain's most vocal AI opponent was asked to lead point-of-sale automation testing. Her eventual endorsement helped sway others who had been hesitant.

Phased Implementation reduces perceived threat. Gradual rollouts allow time for adjustment and build confidence. A logistics company introduced route optimization AI as "suggestions" rather than mandates. This approach eased adoption and minimized pushback.

Two-way Communication builds trust. Listening sessions and anonymous feedback channels surface concerns leadership might miss. Regular forums signal that resistance is acknowledged, not dismissed. Dialogue becomes a tool for alignment.

Success Showcasing makes the case tangible. Demonstrating real-world wins is often more persuasive than abstract arguments. When AI helped sales teams uncover overlooked opportunities, the resulting commission increases spoke louder than any presentation.

Skill Investment signals commitment. Comprehensive training programs show employees they have a future in an AI-augmented organization. This reduces fear and builds confidence. Investment in people is often the most effective form of persuasion.

Middle management requires special attention. They translate strategy into daily reality, influence team morale, and control resource allocation. Yet they often feel vulnerable to AI's potential to flatten hierarchies. Supporting them requires structural inclusion:

- Involve them early in strategy development.
- Define their evolving role in AI-enabled processes.
- Equip them with leadership skills for managing hybrid human-AI teams.

- Align metrics to reward transformation support, not just operational output.
- Create peer forums for sharing challenges and solutions.

Cultural interventions accelerate acceptance. They shape how people interpret change and respond to uncertainty.

Reframe Failure to encourage experimentation. When mistakes are treated as learning opportunities, risk-taking becomes productive. A tech company's "Failure Friday" sessions helped normalize trial and error.

Celebrate Human-AI Collaboration to highlight complementarity. An architecture firm showcased designs where AI generated options and human creativity refined them. The result was a narrative of partnership, not replacement.

Build Communities to support transition. Peer networks (AI user groups, lunch-and-learns, buddy systems) create safe spaces for questions and encouragement. Community fosters resilience.

Share External Examples to reduce isolation. Stories from other organizations facing similar challenges help build confidence. Knowing others have succeeded makes change feel possible.

Resistance to AI adoption is not static; it evolves throughout implementation. In the initial phase, fear and uncertainty dominate. Employees question job security, relevance, and control. As implementation progresses, practical challenges emerge. Users encounter usability issues, integration gaps, and process disruptions. In the later phase, habit and comfort zone dynamics surface. People resist changes that disrupt familiar routines. Post-implementation, resistance shifts again. Requests for fine-tuning and adjustments reflect deeper engagement but also signal lingering friction.

Tracking transformation requires more than anecdotal feedback. Metrics provide visibility into progress and pinpoint areas needing attention. Adoption rates by department and role reveal where uptake is strong and where it lags. Usage frequency and depth show whether tools are being meaningfully integrated. User satisfaction scores offer insight into experience quality. Productivity improvements validate business impact. Innovation indicators such as suggestions and process enhancements signal creative engagement. Support ticket trends highlight recurring pain points and training gaps.

Successful resistance management recognizes that opposition often masks valuable insights. Transform resistance into constructive force. When people feel heard, supported, and included in change, their energy shifts from blocking to improving. The strongest advocates often emerge from initial skeptics who experienced genuine engagement and saw their concerns addressed.

Overcoming initial resistance marks just the beginning. The real test comes after launch excitement fades and AI becomes routine. Without systematic mechanisms for continuous improvement, even successful initiatives atrophy into expensive disappointments.

SUSTAIN SUCCESS THROUGH CONTINUOUS EVOLUTION

AI initiatives require ongoing care, not just successful launches. Many corporate AI projects show early promise, generate enthusiastic press releases, and then quietly fade as attention shifts elsewhere. The result is a growing graveyard of underutilized systems. Only 4 percent of companies achieve substantial value at scale, often because they treat implementation as a destination rather than a continuous journey.

Success measurement begins before deployment and evolves throughout the lifecycle. Without a disciplined approach to tracking progress, organizations risk losing momentum and misjudging impact.

Baseline Metrics establish a foundation for honest comparison. Pre-AI performance benchmarks help distinguish real gains from inflated claims. Too often, organizations report dramatic improvements by comparing against artificially low baselines or cherry-picked time-frames. A credible starting point is essential for meaningful evaluation.

Multi-dimensional Scorecards balance technical, business, and human metrics. Technical excellence alone is insufficient if users avoid the system or if business value remains elusive. Scorecards should reflect adoption, usability, and strategic contribution.

Leading and Lagging Indicators provide a full picture of transformation. Leading indicators such as user engagement and data quality offer early signals of success or trouble. Lagging indicators like cost savings and revenue growth confirm long-term impact. Tracking both helps organizations adjust course before problems escalate.

The shift from launch to habit: By the time the pilot crossed 90 percent accuracy, the leadership team had already started planning the press release. The model triaged service tickets faster than anyone expected. But a month later, the wins felt strangely thin. Adoption lagged in frontline teams. Rework crept back into processes. The dashboard's green bars said "success," yet the operation felt unchanged.

That's when the head of operations reframed the conversation: this isn't a project to finish; it's a habit to build. She borrowed a page from Kaizen: small, steady improvements that compound and moved the center of gravity away from "launch" and toward "learn."

Every Friday, a 30-minute retrospective brought product managers, frontline supervisors, and data scientists into the same room. They didn't debate strategy. They looked at one week's friction. A contact-center agent pointed out that the AI's suggested dispositions were correct but buried behind two extra clicks in the interface. Fixing the layout shaved four seconds off each call; insignificant in isolation, material at scale. The team began to see improvement not as a heroic leap but as a series of unglamorous inches.

Test to learn, not to prove

The team also stopped treating experiments as show trials for the technology. They ran A/B tests that tested ideas, not egos. In one trial, AI recommendations competed head-to-head with human judgment across three queues; in another, two interface variants traded blows; in a third, a revised escalation protocol was pitted against the status quo.

The most valuable finding wasn't that "AI beat humans." It was that the hybrid configuration (AI-first pass, human review on exceptions above a clearly defined risk threshold) delivered the best throughput and the fewest downstream corrections. The team codified that learning as a pattern, not a one-off, and applied it to workflows beyond service tickets.

Energy from the edges

Momentum can't be mandated; it's cultivated. A quarterly innovation challenge gave that momentum a place to gather. The rules were simple: bring a problem from your actual work, assemble a team across functions, and build something real in 48 hours. The prize wasn't cash; it was air cover (time, budget, and stakeholder attention) to carry the idea into production.

One winning team, a trio from compliance and engineering, prototyped a "shadow mode" that ran the model quietly behind human decisions for six weeks. Their insight (documented drift patterns by geography and time of day) rewrote the model monitoring plan. Another team built a lightweight data quality dashboard that became the morning ritual for operators. Leadership didn't predict these ideas; the people closest to the pain invented them.

Benchmarks that raise the bar

Left alone, organizations normalize to their own history. So the team made external benchmarking a routine, not a rescue mission. They compared latency, adoption, and error rates to industry cohorts and to emerging practices in adjacent sectors. The experience was humbling and useful: their accuracy was top quartile, but their cycle times lagged. They weren't failing; they were aiming too low. That external reference point sharpened priorities and upgraded goals.

The first year taught a hard truth: scaling success requires different capabilities than achieving it. Pilots are scrappy; scale is disciplined.

From project to product. The pilot team of brilliant builders handed off to a product organization accountable for roadmaps, SLAs, and support. They established incident management and postmortems. They built observability into the core. A product manager who spoke both operations and data became the single threaded owner. The vocabulary shifted from "model release" to "customer impact."

Geographic expansion isn't copy-paste. When they rolled out to a new region, what worked at home stumbled abroad. Regulations demanded data residency. Customer behavior differed: weekend peaks, mobile-first interactions, WhatsApp expectations. Even label taxonomies needed cultural tuning. The team paused, engaged local leaders, and adapted the design: edge inferencing to meet latency constraints, localized prompts and content, and a retraining pipeline fed by regional feedback loops.

Use-case demand can dilute focus. Success in one domain triggered a flood of requests from every function. They didn't say yes by default; they built a use-case intake with stage gates: problem definition, impact sizing, data readiness, risk assessment, and enablement requirements. A small portfolio council prioritized a handful of bets and explicitly paused the rest. The result was less scatter, more outcomes.

From system to ecosystem. As maturity grew, so did the ecosystem: partner integrations, developer contributions, and data exchanges. The team opened APIs, published standards, and created a "sandbox" for approved experimentation. Governance shifted accordingly: guardrails for privacy and safety, contribution guidelines for extensions, and a lightweight review board to prevent integration sprawl.

Resources that match the moment

Early on, the budget skewed toward technology: infrastructure, licenses, model development. At scale, they rebalanced investment to people, processes, and governance. They funded training and enablement for frontline teams, not just tuning for the model. They resourced risk and compliance from the start, not as after-the-fact approvals. They built documentation, runbooks, and communities of practice: assets that survive reorgs and turnover.

The skill mix evolved too. Builders remained vital, but operators, SRE-minded engineers, product managers, applied researchers, and domain experts became the growth engine. They formalized rotations: data scientists spent stints on the support rota; operators joined quarterly model reviews. Career paths stopped treating "operations" as a cul-de-sac and started recognizing it as a lever for impact.

Vendor relationships matured in parallel. The team moved from transactional procurement to strategic partnerships. They reduced one-way dependence by building internal capabilities, negotiated for transparency on roadmaps and model cards, and set shared metrics tied to outcomes. Where it made sense, they co-developed features with partners; where it didn't, they insourced critical competencies.

How momentum is sustained

By year two, the organization looked different from the outside and felt different from the inside. Not because of a single breakthrough, but because improvement had become a habit. A few practices made the difference:

Leadership stayed after the launch. The executive sponsor didn't vanish once the pilot went live. She chaired monthly reviews that focused on user outcomes, risks, and learning - not slide count. Her presence protected capacity for the unglamorous work that makes systems reliable.

Improvement moved into the routine. The Friday retrospective wasn't optional. The data quality dashboard opened every morning stand-up. Small wins earned visible celebration: an engineer's one-line fix that reduced latency, an agent's insight that improved the deflection script.

Stability and innovation were managed as a portfolio. The team tracked health metrics (uptime, latency, override rates) alongside learning metrics (experiments run, time to decision, adoption curves). When stability dipped, they slowed expansion. When headroom appeared, they explored.

Capability building was continuous. Training wasn't a one-time onboarding. New hires learned how models fail as well as how they succeed. Teams practiced incident response like fire drills. Documentation stayed living, not dead-on-arrival.

Failure taught without blame. Postmortems were blameless and specific. A misrouted escalation didn't trigger finger-pointing; it led to better guardrails and clearer ownership. People raised issues early because they trusted the system.

The payoff

Outcomes followed the behavior. Cycle times fell not in a headline-grabbing week, but month after month. Override rates dropped as the hybrid pattern matured. Adoption climbed because the system respected the realities of frontline work. New markets came online faster because the team had a playbook for localization. Even the vendor ecosystem became a source of advantage rather than a dependency.

The lesson is simple to say and hard to practice: treat improvement as a daily discipline, treat scaling as a different phase of work, and evolve your resourcing as you mature. Do that consistently, and progress compounds and endures.

The journey doesn't end at launch. Technology keeps advancing. Business priorities shift. Competitive landscapes redraw themselves overnight. Organizations that embed **continuous evolution** into their DNA build platforms for ongoing innovation.

AI transformation thrives on this duality: the spectacular and the systematic. Breakthrough moments matter, but so does the quiet discipline of maintenance. Models need tuning. Processes need refinement. Guardrails need reinforcement. Sustained success comes from honoring both: the headline-grabbing leap and the unglamorous inch.

SUMMARY

The frameworks presented here work as an integrated system. Each component reinforces the others: honest assessment informs strategy, strategy guides leadership action, leadership enables workforce development, skills support ethical implementation, ethics addresses resistance, and reduced resistance enables continuous improvement.

The journey isn't linear. You will encounter setbacks, discover blind spots, and face unexpected challenges. These frameworks provide recovery mechanisms and adaptation pathways, not guaranteed success formulas. Context matters: company size, industry regulations, and cultural factors all influence implementation.

Three principles can be applied to these frameworks:

- Start with brutal honesty about current capabilities
- Maintain sustained leadership commitment through inevitable difficulties
- Focus relentlessly on human factors that determine adoption

Organizations that master these frameworks develop systematic transformation competencies that prepare them for whatever disruptions emerge next. The future belongs to enterprises that seamlessly blend human judgment with machine intelligence.

Choose your starting point based on your most critical gap. Take action. Build momentum through small wins. Learn and adapt. The path forward requires courage, persistence, and systematic execution. With these frameworks as guides, you're equipped for the journey.

Chapter takeaways:

- Sustainable AI transformation begins with a rigorous readiness assessment: aligning strategy, structure, systems, and culture to support change.
- Leadership must move beyond ceremonial support to active sponsorship, resource allocation, and visible engagement throughout the journey.
- Universal upskilling and AI literacy are essential at every level, not just among technical teams.
- Ethical governance frameworks must be operationalized, not just stated, to build trust and manage risk.
- Resistance is expected and should be transformed into momentum by involving skeptics in solution design and celebrating early wins.
- Continuous improvement is institutionalized through feedback loops, performance measurement, and regular adaptation of processes and resource allocation.
- Scaling success requires shifting from pilot projects to enterprise-wide adoption, with investment rebalancing from

technology to people, process, and governance as maturity grows.
- The journey is ongoing: organizations that sustain momentum embed learning, adaptability, and human-centric values into their operating model.

With the foundational frameworks for execution now in place, the focus shifts to the nuances that separate functional implementation from market dominance. As organizations move past the initial hurdles of adoption, they encounter a new set of sophisticated challenges: navigating a fragmented global regulatory landscape, managing high-stakes ethical risks at scale, and preparing for technologies that have yet to be invented. Chapter 8 explores these advanced frontiers, equipping leaders to navigate the complexities of an AI-first world where the rules are constantly being rewritten.

8

ADVANCED TOPICS IN AI-DRIVEN ORGANIZATIONAL CHANGE

The companies that thrive will be those that navigate complex regulations, build robust governance, operationalize ethics, reimagine their workforce, sustain momentum, manage risks, and evolve their leadership for an AI-first world.

When AI moves from pilot to production, from experiment to enterprise standard, new challenges emerge that test even seasoned executives. These advanced topics determine whether your AI initiatives become competitive advantages or costly failures. This chapter equips you with the frameworks and strategies to excel where others stumble.

CLAIM YOUR STRATEGIC POSITION IN THE REGULATORY MAZE

Smart companies transform compliance from burden to advantage by aligning internal practices with the toughest standards, building trust while competitors scramble.

The regulatory landscape for AI resembles a complex chess game played simultaneously on multiple boards. The European Union leads with its comprehensive AI Act, the world's first framework specifically targeting artificial intelligence. This landmark legislation takes a risk-based approach, classifying AI into four tiers: minimal risk, limited risk, high risk, and unacceptable risk. Medical diagnostics and hiring algorithms fall into the high-risk category, requiring rigorous assessments and human oversight, while spam filters occupy the minimal risk tier with few requirements.

Most EU AI Act violations carry fines up to €15 million or 3 percent of global revenue. The worst offenses (such as deploying prohibited "unacceptable risk" AI systems) can trigger penalties reaching €35 million or 7 percent of worldwide turnover. With enforcement looming as EU member states finalize provisions in mid-2025, companies operating in or selling into the EU must prepare now.

The United States presents a different challenge: a complex web of guidelines and sectoral regulations rather than unified federal law. The Equal Employment Opportunity Commission warns against biased AI in hiring. The FTC targets deceptive AI practices. Financial regulators expect AI models to comply with fair lending and model risk management rules like the Federal Reserve's SR 11-7. New York City already mandates bias audits for automated hiring tools. While federal momentum builds (including the 2024 Executive Order on AI and signing the Council of Europe's AI Framework Convention) companies navigate a patchwork of state laws and agency guidance.

Other regions reveal diverse regulatory philosophies:

China implemented strict AI governance in 2022-2023, introducing rules for recommendation algorithms and generative AI that emphasize national security and social stability. Companies must register algorithmic services and comply with content guidelines.

Singapore and Japan favor voluntary frameworks. Singapore's IMDA published Model AI Governance Frameworks and runs an AI Verify program for voluntary testing. Japan issued guiding principles that signal expected practices without legal mandate.

United Arab Emirates takes a sectoral approach, requiring transparency for AI in healthcare and finance while leaving other sectors less regulated.

India currently relies on existing IT and privacy laws while drafting a Digital India Act that may address AI specifically.

Forward-thinking organizations adopt a "comply to the toughest standard" approach. Like GDPR reshaped global privacy practices, the EU AI Act's influence will likely extend worldwide. Companies aligning with its requirements (e.g., risk assessments, data quality controls, documentation, human oversight) create foundations that meet or exceed requirements elsewhere.

Industry-specific regulations add complexity. Banks must integrate AI with existing model risk management frameworks. Healthcare organizations navigate FDA evaluation of AI/ML medical devices. Insurance regulators scrutinize AI underwriting for bias. Each sector demands specialized compliance approaches.

Success requires systematic action:

Map your AI landscape: Create a comprehensive inventory of AI systems, documenting their purpose, data sources, and risk levels

Establish AI compliance functions: Designate officers or teams within risk or legal departments to track regulatory developments globally

Leverage international standards: OECD AI Principles, endorsed by over 40 countries, provide consistent baseline practices

Align with emerging frameworks: ISO/IEC 42001, an AI management system standard under development, may offer certifiable compliance pathways

Compliance with AI laws builds customer trust and avoids costly penalties. Companies adept at navigating regulations gain competitive advantage as customers, partners, and investors increasingly favor organizations demonstrating responsible AI practices. As regulatory expectations rise, organizations that treat compliance as a foundation position themselves to build deeper trust and unlock strategic advantage.

ARCHITECT GOVERNANCE THAT SCALES

Organizations need structures that ensure accountability, manage risk, and enable innovation while maintaining the agility to adapt as AI capabilities evolve.

Governance is how we humanize the swarm at scale. It is the mechanism by which we encode human values like fairness, privacy, and dignity into the 'black box.' Without these guardrails, the swarm optimizes blindly. With them, it becomes a steward of your organizational ethics.

While 80 percent of organizations now dedicate risk management resources to AI according to IBM Institute for Business Value research, few have built comprehensive frameworks that balance control with innovation. The most effective governance models share key characteristics that separate leaders from laggards.

Clear ownership and accountability defines success. The CEO and C-suite maintain ultimate accountability for sound AI governance, supported by legal counsel for regulatory compliance and auditors for validation. Establish cross-functional AI governance committees drawing from IT, data science, risk management, legal, compliance, and business units. Some organizations appoint Chief AI Officers (CAIO) or expand Chief Data Officer mandates to cover AI governance.

Internal AI ethics boards add crucial perspective. Leading IT companies have published responsible AI principles with internal review processes to enforce them. These boards often include external advisors or independent experts, bringing unbiased viewpoints to complex ethical questions.

Comprehensive AI inventories provide essential visibility. According to published reports OneTrust identifies compiling an AI inventory as "one of the first critical steps in starting an AI governance program." This centralized register documents every AI system's purpose, data sources, algorithms, risk assessments, and responsible parties. High-risk applications like medical diagnoses or hiring recommendations trigger additional governance steps. Documentation standards requiring model "datasheets" or "model cards" feed into the inventory, ensuring institutional memory about how systems were built and their intended use.

Risk management processes must be systematic. The U.S. NIST AI Risk Management Framework, released in 2023, provides a voluntary standard built around four core functions:

Govern: Establish oversight and accountability structures

Map: Identify contexts and risks of AI systems

Measure: Analyze and assess risks through bias testing and security evaluations

Manage: Apply controls and strategies to mitigate identified risks

Under this framework, teams conduct risk assessments at project inception, asking critical questions: Does this use case pose privacy risks? Could it affect someone's rights or safety? High-risk projects require formal approval from governance committees. Controls like bias testing, validation, and human-in-the-loop checkpoints mitigate identified risks.

Crucially, governance doesn't stop at launch. NIST and others emphasize that trustworthy AI demands continuous monitoring and improvement, not a one-time approach. Procedures must detect when AI outcomes drift into risky territory, triggering retraining or adjustment.

Accountability and audit mechanisms ensure compliance. Governance models define who's accountable for each AI system's outcomes, often assigning "model owners" or "stewards" from business units deploying the AI. Internal auditors or specialized AI audit teams validate that systems meet organizational standards. This includes technical audits (checking for bias, robustness, security) and process audits (ensuring teams followed mandated steps).

Many mature organizations require model validation before production deployment, echoing how banks validate credit risk models. Automation supports oversight: software can track model performance and alert when metrics exceed thresholds, while tools generate explainability reports to aid auditing.

Training and culture amplify governance effectiveness. IBM's guidance emphasizes that "every leader and every employee has a part to play in AI governance; it's a collective responsibility." Best practices include mandatory training on AI ethics and governance principles, similar to data privacy or cybersecurity education. Microsoft developed extensive responsible AI training for engineering staff. Some companies implement "champion" programs identifying ethics advocates within teams.

Leaders must foster environments where employees feel safe raising AI concerns. When someone spots potential unfairness in a model, governance culture ensures they know where to escalate and trust serious action will follow.

Organizational structures must match maturity levels. Companies initially favor centralized models: a central AI governance team sets policies and reviews major initiatives. This ensures consistency but may create bottlenecks as AI use grows. Many evolve toward federated models, establishing common standards centrally while delegating day-to-day governance to business units or regions.

For example, a bank might maintain central AI governance policy while each division (retail banking, insurance) operates its own model risk committee applying that policy. Federated models work well in large enterprises by distributing expertise while maintaining coordination. Most firms start centralized, then evolve toward hybrid approaches as governance matures.

A 2024 AI governance report reveals professionals face ongoing challenges: integrating governance into agile development without stifling innovation, handling third-party AI tools entering the enterprise, and scaling oversight when hundreds of applications operate simultaneously. Success requires sharing tools and approaches from checklists for project managers to templates for ethical impact assessments.

The payoff justifies the investment. Consider deploying an AI-powered customer service chatbot. Without governance, businesses might launch quickly to cut costs, only to face backlash when the bot gives harmful advice or violates privacy. With governance, the project undergoes review: Does it adhere to data handling rules? Has it been tested for inappropriate responses? Is there human escalation for unhappy customers? Addressing these questions dramatically reduces post-deployment failures.

Organizations with mature AI governance innovate with confidence. They pursue ambitious projects knowing guardrails manage risks. Stakeholders trust companies taking responsible steps. As one executive noted, effective governance allows organizations to "move fast without breaking things," accelerating AI adoption while maintaining control.

SCALE ETHICS WITHOUT COMPROMISE

Companies that systematically prevent bias, ensure transparency, and maintain human oversight create sustainable competitive advantages while those that don't face reputational ruin.

The ethical challenges of AI multiply exponentially with scale. A healthcare algorithm affecting millions revealed how indirect bias creeps in: research published in 2019 found an algorithm used on over 200 million patients was heavily biased against Black patients. The AI predicted who would benefit from extra care for chronic conditions, but it used healthcare costs as a proxy for health needs. Because historically Black patients incurred lower healthcare costs due to unequal access to care, the algorithm systematically assigned lower risk scores to Black patients with the same conditions as white patients.

Once discovered, researchers worked with the developer Optum to adjust the algorithm, reducing bias by 80 percent. But imagine if it hadn't been caught - millions could have received substandard care due to an AI's unethical design.

Real-world AI failures provide stark lessons across industries:

Criminal justice systems show algorithmic injustice. According to public reporting, the COMPAS algorithm for predicting recidivism became notorious after a 2016 ProPublica investigation found it was far more likely to incorrectly label Black defendants as high risk: 45 percent false positive rate for Black defendants versus 23 percent for white defendants. This resulted in Black individuals being disproportionately flagged as dangerous and potentially receiving harsher treatment. The bias was attributed to data and design choices that reflected existing inequalities in arrests and sentencing.

Recruitment AI can entrench discrimination. According to public reports, Amazon's AI hiring tool, disclosed in 2018, demonstrated discrimination against female candidates. The system was trained on ten years of resumes—most of which came from men due to the male-dominated nature of the tech workforce. As a result, the AI learned to favor resumes that resembled those of previous successful (male) applicants, and it penalized resumes mentioning women's clubs or women's colleges. After the bias was identified, Amazon discontinued the tool. This case shows how bias can enter AI systems unintentionally when training data reflects existing inequalities.

These examples highlight that AI can and will scale biases or ethical problems if not managed. When deployed at scale, an AI impacts hundreds of thousands or millions of people. The stakes are ethical lapses that harm customers, invite lawsuits, violate anti-discrimination laws, and severely damage reputation, and demand systematic approaches.

Build ethics into design from day one. Adopt "ethics by design" like privacy by design in software. At project inception, conduct algorithmic impact assessments to identify ethical and social risks before building. Teams ask hard questions: What negative impacts could this system have? Who might be adversely affected? Are there biases in our training data? These assessments, which some regulators may mandate for high-risk AI, force teams to consider harms and mitigations early.

Ensure diverse and representative data. Many biases stem from data. Training data must represent the real population as diversely as possible. Facial recognition systems that misidentify people with darker skin tones reflect training images lacking balanced representation of genders, ethnicities, and skin tones. When historical data reflects unequal treatment, teams must supplement or rebalance datasets to reflect fair outcomes rather than biased history. Sometimes new data collection fills gaps; other times, techniques like data augmentation create more balanced datasets.

Test rigorously and systematically for bias. Throughout development, data scientists must measure model outcomes across demographic groups (race, gender, age, as appropriate and lawful) to identify disparities. If a loan approval AI shows significantly lower approval rates for one ethnic group even controlling for credit factors, that signals bias requiring attention.

According to public reporting, IBM's AI Fairness 360 toolkit and Microsoft's Fairlearn are open-source libraries designed to evaluate and mitigate bias in datasets and models. When bias is detected, public reports indicate that techniques such as reweighting training data, excluding problematic variables, or adjusting decision thresholds are often used to improve fairness. Importantly, bias testing continues even after deployment, as models encounter new data.

Demand algorithmic transparency and explainability. Organizations should strive for transparency in how algorithms operate. While not every complex model can be fully explainable, companies should at least be transparent about AI use and provide explanations where feasible.

Techniques under Explainable AI (XAI) can assist with this: methods like Local Interpretable Model-agnostic Explanations (LIME) or SHapley Additive exPlanations (SHAP) highlight which features influenced particular decisions. Some companies produce explainability reports; for example, a fintech might inform users, "Your credit was denied; key factors were income below $X and a recent delinquency," derived from model inputs.

Best practice includes transparency statements about AI systems. Businesses using AI chatbots should disclose the interaction is AI-driven. Honesty about AI involvement sets expectations and builds trust rather than attempting deception.

Maintain human oversight for high-stakes decisions. As AI scales, complete automation tempts organizations. But ethical management often requires human-in-the-loop oversight, especially for high-impact decisions. This might mean final approval by humans (as banks do for AI-flagged fraud cases) or AI operating in advisory mode, giving recommendations humans consider.

Human oversight provides sanity checks and catches ethical issues algorithms miss. A hospital using AI to prioritize patients might have doctors review and adjust recommendations based on factors AI couldn't know. Organizations must clearly define when AI operates autonomously versus when human review is require, usually based on risk level. The higher the stakes for individuals, the more oversight needed.

Critically, organizations should empower employees to challenge or override AI when necessary. If customer service agents feel an AI suggestion inappropriate for particular clients, the culture should support their judgment to deviate rather than blindly follow machines.

Foster ethical culture through training and incentives. Just as governance requires cultural support, so does ethics. Company AI ethics guidelines should be communicated and made concrete for teams. IBM developed extensive responsible AI training for engineering staff. Microsoft created "champion" programs identifying and empowering ethics advocates within teams.

Running workshops or scenario training proves effective: pose dilemmas like "Our facial recognition system misidentifies certain minorities at higher rates: what should we do?" and let teams discuss. Checklists for product managers cover ethical questions at each development phase. By institutionalizing these practices, organizations reduce reliance on individual heroes and everyone becomes a steward of AI ethics.

Monitor continuously and audit for ethics. Ethical management doesn't end at deployment. AI models can "drift", meaning their performance and fairness degrade if data entering the system changes over time. Therefore, continuous monitoring and auditing is crucial.

As a best practice, periodically audit algorithms in production for outcomes and potential biases. A hiring algorithm might receive audits every six months to ensure it's not developing new biases as job market data shifts. Problems trigger model retraining or other interventions.

Monitoring can be partly automated using dashboards tracking key metrics by demographic segments alert if significant disparities emerge. Additionally, establishing feedback channels encourages users or employees to report odd or unfair AI outputs. Some organizations implement "AI incident" reporting processes internally, akin to software bug or safety incident reporting.

Engage external perspectives for objectivity. At scale, involving external experts to review AI systems from ethical standpoints helps catch insider blind spots. This could include academic partnerships (many universities have AI ethics programs interested in real-world cases), third-party audits or certifications, or participation in industry consortia setting ethical standards like the Partnership on AI.

By embracing external transparency (e.g., publishing fairness test results or steps taken to ensure ethical AI) organizations signal accountability. Some companies allow external observers to evaluate algorithms with appropriate confidentiality, which, while potentially uncomfortable, catches issues and enhances credibility.

Eliminating biases often improves overall accuracy since models aren't swayed by irrelevant factors. Adding transparency reduces errors because humans spot when AI reasoning goes awry. Considering user perspectives creates AI services people actually trust and want to use. Ethics and business value reinforce each other long-term.

Moreover, regulators increasingly expect these ethical controls. The EU AI Act explicitly requires measures for data governance and bias monitoring in high-risk AI. Early compliance prepares companies while positioning them as industry leaders.

An enterprise successfully managing AI ethics at scale enjoys what might be called "trust capital", meaning customers, employees, and partners have confidence the organization uses AI wisely. But this trust is fragile; building it takes far more effort than losing it through scandal. To preserve this trust capital and harness AI responsibly, organizations must also reimagine their workforce by aligning talent strategies with the realities of human-AI collaboration.

TRANSFORM YOUR WORKFORCE FOR THE AI ERA

Companies that reskill proactively and redesign work thoughtfully will unlock productivity gains while those that don't will face talent crises.

According to public reporting, such as IBM's Global AI Adoption Index, approximately half of business leaders worldwide anticipate that at least 50 percent of their workforce will require upskilling in the near term. Additionally, LinkedIn's published data indicates that more than one in ten job postings today are for roles that did not exist just a couple of decades ago. Looking forward, publicly cited projections estimate that by 2030, about 70 percent of the skills required for jobs will have changed.

For leadership, these statistics demand a strategic approach to workforce planning. Traditional models focusing on hiring and succession for known roles must now account for automation of tasks and creation of hybrid roles. By breaking jobs into component tasks, companies can identify which can be delegated to AI versus which require distinctly human strengths.

Anticipate job evolution rather than elimination. Most roles will transform. AI "takes over the tedious work," freeing employees to focus on higher-value activities. AI handles routine customer inquiries via chatbot while human service reps manage complex cases and sensitive interactions. AI generates basic analytical reports, allowing human analysts more time interpreting results and advising strategy.

Execute reskilling at unprecedented scale. Leading companies invest heavily in workforce development. Some approaches include:

Internal AI academies: Some organizations have launched large-scale reskilling programs, providing employees with courses in data science, machine learning, and other technology fields through collaborations with online learning platforms and universities.

On-the-job training: Companies embed AI mentors or use AI-based training tools that adapt to each employee's progress

Job rotation programs: Marketing managers spend rotations with data analytics teams to learn how AI-driven customer insights are generated, building analytical skillsets

Tuition reimbursement: Funding external certifications in AI, data analytics, or related fields shows workforce investment in growth

Strategic segmentation identifies roles most affected by AI and most critical to future success; these become the focus of intensive upskilling. For soon-to-be-phased-out roles, companies provide retraining for new positions or transition support.

Create entirely new AI-augmented roles. Organizations incorporate emerging positions into org charts:

AI System Trainers/Explainers: Help train AI models and translate AI outputs for business teams. "AI business translators" understand both data science and business needs, ensuring AI projects solve right problems.

AI Ethics Officers: Ensure AI is used ethically, conducting impact assessments, and often sitting in risk or compliance teams.

Data Curators: Manage data feeding AI, ensuring quality, completeness, and lack of bias.

Automation Coaches: Help departments find and implement automation opportunities, serving as internal AI adoption consultants.

Strategic workforce planning means determining which roles your organization needs, then deciding whether to build (train existing staff) or buy (hire new talent). Most companies pursue a mix:

upskilling insiders who know the business while hiring external experts to infuse new technical capabilities.

Address employee fears through transparency. According to publicly reported findings from a recent ADP global survey, 31 percent of workers expressed some level of fear that AI could potentially replace their jobs. Simultaneously, 50 percent see AI's potential for positive job impact, revealing a contradiction of excitement and fear common across organizations. When rolling out AI, explain why and how it's being introduced, what will change and what won't. If the plan uses AI to assist rather than replace, state that clearly and back it with actions like retraining commitments.

Involve employees in AI initiatives through pilot program volunteers or focus groups gathering input. When people own transformation pieces, they're less likely seeing it as imposed threat. Research indicates employees' openness to AI strongly correlates with leadership trust. Workers expressing high trust in senior leaders are far more likely welcoming AI advances.

Highlight relatable AI benefits in daily work. For example, showing office workers how AI summarizes lengthy emails transforms AI from threat to assistant. Emphasizing AI removes drudgery enabling more meaningful work can turn wary employees into AI advocates. When educated on personal benefits, many become excited about shedding boring tasks.

Adapt organizational culture and structure. AI integration pushes organizations toward more collaborative, cross-functional cultures. Lines between technical and business teams blur when AI projects require data scientists needing domain input from business experts, while business units need technical support implementing AI.

Progressive companies break down silos, creating agile teams mixing IT, AI experts, and business stakeholders to co-create solutions. Establish AI Centers of Excellence (CoEs) with small expert teams to serve the organization with expertise, tools, and governance. These

COEs provide training grounds where business staff are seconded for skill development before rotating back as "AI ambassadors".

Evolve performance management and incentives. If AI performs part of an employee's work, measuring that employee's performance requires adjustment. Focus might shift to uniquely human contributions like creativity, client relationships, or AI accuracy oversight.

Additionally, encourage and reward employees learning new skills and improving processes with AI. Some firms explicitly include goals like "Identify two opportunities to use automation/AI to improve your team's workflow" in management performance plans. This creates bottom-up AI adoption drive while reinforcing that using AI positively contributes rather than threatens jobs.

Handle displacement with humanity. Even with extensive reskilling, certain roles may reduce. Strategic workforce planning should include humane transition plans for displaced workers. Best-case scenarios involve reskilling into other company roles. When that's infeasible, providing support through severance, job placement assistance, and external opportunity retraining demonstrates socially responsible leadership.

The internal narrative should emphasize company values for its people: AI adoption aims to elevate work and competitiveness, not simply cut employee costs. Organizations known for treating employees well during tech transitions sustain higher morale and reputation, helping retain talent needed to run AI systems.

Design human-AI collaboration workflows. Practical workforce integration requires determining workflows between humans and AI. Experiment with "human-AI teams" where AI outputs receive human review or enhancement. In content moderation, AI algorithms might flag 90 percent of potentially problematic posts, with human moderators focusing time reviewing those flags. Team workflows and training must establish how humans understand AI capabilities and limitations. Workers blindly trusting AI outputs allow errors to slip through; those distrusting everything lose efficiency gains. Training should cover interpreting AI recommendations and calibrating trust levels.

This interplay can increase job satisfaction when done right; employees apply judgment and expertise rather than slogging through repetitive tasks. Early studies on human-AI collaboration show mixed results: it improves outcomes when well-designed but worsens them if not.

Cultivate growth mindset across leadership. Perhaps most importantly, leaders must foster growth mindset throughout the workforce, where continuous learning is valued. In AI-driven environments, job roles may evolve yearly with new tools; employees need the mindset of "I'm not a know-it-all, I'm a learn-it-all" for adaptability. By encouraging learn-it-all culture, Microsoft upskilled its massive workforce for cloud and AI innovations. Leaders at all levels should model this by actively learning AI themselves and transparently sharing their learning journey.

The payoff for successful workforce transformation extends beyond efficiency. When mundane work automates, employees focus on creative, strategic, and interpersonal work aspects often bringing greater fulfillment. Strong skill development programs show company investment in growth, boosting morale and retention.

According to publicly reported findings from a 2024 SHRM study, three out of four HR professionals indicated that as AI automates technical tasks, human intelligence and uniquely human skills are becoming increasingly vital in the workplace. Public sources also quote a CEO stating, "Human capital is more important than ever" in the age of AI, emphasizing that people are essential for guiding implementation and contributing creativity and empathy—qualities machines cannot replicate.

By identifying skill needs, reskilling proactively, managing change empathetically, and redesigning work, leaders ensure their workforce remains an asset, not casualty, of AI dominance.

BUILD AI PROGRAMS THAT ENDURE

Only a quarter of companies generate sustained value from AI at scale. The difference between success and failure lies in treating AI as organizational capability, not isolated projects.

According to publicly reported findings from a September 2025 MIT study, The GenAI Divide: State of AI in Business 2025, there remains a persistent failure rate for pilot programs. The study reports that an overwhelming 95 percent of enterprise AI pilots do not deliver measurable financial returns, largely because organizations face challenges in translating individual productivity improvements into scalable business value.

The path from AI pilot to enterprise value is littered with abandoned projects, a phenomenon known as "pilot purgatory" or the "AI plateau" where progress stalls after initial excitement. Many companies treat AI like a middle school science fair volcano: it's fun to watch it fizz for five minutes, but nobody suggests we should use it to power the building. To escape pilot purgatory, you must stop treating AI as a novelty act. You need to connect the 'cool tech' to the 'boring ledger.' If it doesn't save money or make money, it's just digital vinegar and baking soda. A global Qlik survey identified the key factors hindering successful AI deployment: lack of AI skills in the workforce, governance issues, and resource constraints, leaving many projects stuck in planning phases.

Understanding why AI initiatives lose momentum reveals the path to sustainability:

Skills shortage: Initial projects driven by small expert teams or external vendors hit scaling ceilings without broader AI capabilities

Infrastructure gaps: Bespoke data wrangling works for experiments but production requires robust data pipelines and computing infrastructure

Governance deficiencies: Without proper oversight, AI proliferation leads to compliance violations that trigger leadership pullbacks

Unclear ROI: Executives asking "What are we getting from this?" a year later find enthusiasm and funding drying up without clear metrics

Change fatigue: Employees experiencing continuous AI implementations may resist if benefits aren't apparent

Maintenance neglect: Treating AI as "set and forget" leads to model drift and degrading performance

Leadership turnover: When executive sponsors leave or priorities shift, AI projects lose champions

Success requires systematic approaches across multiple dimensions:

Demonstrate ROI relentlessly from day one. Link AI projects to measurable business outcomes and set intermediate milestones: cost savings, revenue upticks, efficiency gains, customer satisfaction improvements. Communicate these wins internally to maintain momentum.

According to publicly reported findings from LinkedIn, 51 percent of small and medium-sized businesses (SMBs) using Generative AI experienced a 10 percent increase in revenue. By regularly updating business cases with new data on realized benefits, organizations can create a reinforcing cycle where each success drives further funding and expansion, leading to continued growth.

Embed AI into core operations. Long-term sustainability comes from making AI integral to organizational functions. Redesign workflows and business processes to include AI where it adds value. Update standard operating procedures to incorporate AI tools. When

AI becomes entwined in daily operations rather than bolt-on experiments, it becomes indispensable.

Incorporate AI into product roadmaps and customer offerings. Create strategic roadmaps specifying which product lines or operations will be "AI-powered" with defined capabilities by target dates. This creates organizational alignment that AI is embedded in the future.

Invest in scalable MLOps infrastructure. Adopting modern Machine Learning Operations platforms proves key to sustaining AI. MLOps provides automated, reliable pipelines for data, model training, deployment, and monitoring - analogous to how DevOps transformed software development.

Implementing MLOps reduces manual effort maintaining models. Model monitoring systems automatically alert when accuracy drops or data drift is detected, prompting retraining. Scalable cloud infrastructure ensures AI systems handle usage growth without performance issues.

High-performing companies often build internal "AI platforms", suites of tools and environments any team can use to develop and deploy AI, with security and compliance built-in. This prevents teams reinventing wheels and accelerates new project scaling. Treat AI models like critical software assets: with version control, testing, deployment pipelines, and maintenance schedules.

Foster continuous innovation culture. Long-term AI success requires continuously improving existing models and expanding to new use cases. Encourage a mindset that AI systems are always in beta, open to improvement. Establish regular review cycles (quarterly, annually) assessing whether key AI systems still meet needs or if new techniques could improve performance.

Technology evolves rapidly: what was cutting-edge two years ago will be surpassed by new algorithms or larger datasets. Monitor AI research and be ready to upgrade models or approaches. Maintain

R&D groups or academic partnerships to stay at forefront, feeding innovation back into operational systems.

Design for scalability from inception. Overcome the pilot-to-production gap by designing projects with scaling in mind from the start. Choose pilots that, if successful, can roll out enterprise-wide relatively easily. Develop repeatable frameworks; if a bank creates AI for credit card fraud detection, that framework might extend to online banking or insurance claims.

The more you templatize solutions, the easier scaling becomes. Consider phased change management: involve representatives from different business units in pilot design so they become champions driving adoption in their units later. Scale in phases to manage risk and incorporate feedback. Once regional or departmental pilots are successful, expand to new areas rather than attempting a big-bang rollout.

Maintain executive and board engagement consistently. Sustained AI programs benefit from continued top-level sponsorship. Add AI progress as regular agenda items in leadership and board meetings. Some companies form AI steering committees at board level or include AI oversight in risk committee charters.

Executives should receive AI education to steer strategy effectively, understanding the implications of the latest Gen-AI waves for business models, for instance. When the C-suite remains actively interested (i.e., asking for project updates, inquiring how AI solves strategic problems, celebrating AI-driven wins) it signals longevity of commitment.

Set realistic timelines to avoid AI hype fatigue. Communicate that AI represents long-term transformation, not a quick fix. This tempered, steady leadership support prevents organizations from cycling through boom-bust hype phases.

Address trust and risk proactively. Often AI initiatives pause or reverse due to scares or incidents: AI errors causing costly mistakes or bias concerns arising. Implementing strong risk management and ethics practices helps avoid crises souring leadership or public opinion on AI.

Building track records of responsible AI usage creates organizational resilience. When issues occur, stakeholders trust the company can fix them and continue rather than abandoning AI altogether. Transparency about AI performance helps; if employees and leaders know system strengths and limitations, they're less disillusioned when occasional errors occur.

Plan financially for long haul. Budget for ongoing AI investment beyond initial project costs: maintenance, future upgrades, and critically, continuous talent investment. Create central "AI funds" or budget pools supporting promising initiatives across departments, preventing immediate P&L ROI requirements from killing projects with enterprise value.

View AI capability as capital investment with multi-year payback rather than expecting immediate quarterly returns. A retailer might not see full ROI from AI-driven supply chain optimization for two to three years as refinement occurs, but eventual efficiency gains could prove massive. Avoid cutting AI initiatives prematurely due to short-term pressures, a common "pilot purgatory" cause occurs when modest initial results prompt funding withdrawal before models mature.

Spotlight enduring wins. Inside the company, amplify AI programs that are running reliably and expanding. Visible success boosts morale and momentum. If a recommendation engine piloted in one country is now deployed globally and lifting conversion, turn it into a case study. Show the journey, not just the outcome: what made it work, how the team scaled it, and what others can reuse. This is how isolated pilots become institutional capability.

Externally, disciplined storytelling builds stakeholder confidence. Publish what you can: blog posts on scaling patterns, talks on operating models, and lessons on governance and risk. The point isn't hype; it's credibility. Companies that share pragmatic playbooks are seen as innovators and signal to investors and partners that they can execute at scale, creating healthy internal pressure to keep raising the bar.

Build the flywheel. Early wins attract investment. Investment upgrades data, platforms, and skills. Better foundations enable bigger wins. Over time, AI stops living in a project portfolio and becomes part of the operating system. You know you've arrived when people no longer say "AI project"; they just say "the pricing program," "the collections process," or "the supply plan," with AI embedded by default.

Make sustainability nonnegotiable. As AI scales, so do compute and energy demands. Pair ambition with efficiency: set targets for carbon-aware workloads, optimize models and code paths, and favor cloud regions and architectures that reduce emissions. Societal sustainability matters too. Ensure benefits are inclusive, jobs are redesigned with workers, and communities aren't left behind. Tie these choices to corporate responsibility commitments; they earn long-term stakeholder support and de-risk the roadmap.

Done right, AI becomes a durable engine of innovation and productivity. Organizations that invest through the growing pains build the capabilities that future-proof them in an AI-driven economy. Even so, the journey isn't linear. The next challenge is sophisticated risk management: governance, controls, and assurance that keep pace with scale and impact.

MASTER CRISIS BEFORE IT STRIKES

Organizations with robust risk management, tested response playbooks, and transparent communication turn potential disasters into demonstrations of competence.

AI introduces novel risks that can escalate with frightening speed. The spectrum of AI-related risks spans operational failures to reputational catastrophes, each requiring specific preparation:

According to public reports, the 2012 Knight Capital incident stands as a notable warning: a glitch in algorithmic trading software led the firm to lose $440 million within minutes. Publicly available accounts indicate that Knight Capital lost 75 percent of its value overnight and was subsequently sold. This widely reported event highlights how automation failures can devastate established companies more rapidly than traditional, human-driven crises.

According to public reports, Microsoft's Tay chatbot in 2016 began tweeting offensive content within hours of launch after internet users exploited its learning algorithms. The company subsequently shut down the bot and issued public apologies. This widely reported case illustrates how, for customer-facing AI, a single viral incident can undermine years of brand building.

AI systems operating as designed can still violate norms or regulations. A recruiting AI that discriminates could trigger lawsuits under employment law. A credit model inadvertently violating fair lending laws could bring regulatory action. These incidents combine compliance failures with PR crises.

As AI becomes core to operations, it attracts sophisticated adversaries. Threat vectors include:

Data poisoning: Hackers subtly manipulate training data to corrupt model outputs

Adversarial inputs: Specially crafted inputs fool AI systems: researchers showed placing stickers on stop signs could cause self-driving car vision systems to misclassify them

Model theft: Stealing proprietary models representing significant intellectual property

System manipulation: Attackers causing financial AI to generate fraudulent trades combine cyber and financial risks

The 2010 Flash Crash demonstrated how multiple firms' trading algorithms without proper circuit breakers caused the Dow to plunge nearly 1,000 points in minutes before rebounding. This shook market confidence globally. Similarly, AI systems dynamically setting prices might accidentally collude with competitors' AI, creating antitrust violations.

Deepfakes and AI-generated misinformation create new vulnerabilities. A deepfake video of a CEO making inflammatory statements could tank stock prices before being debunked. Companies must prepare for AI-driven misinformation campaigns against them - a novel PR challenge.

According to tracking data, the number of AI incidents reported has been rising approximately 30 percent year over year. Being prepared is a competitive necessity.

Identify and assess AI risks comprehensively upfront. During AI system development, conduct thorough risk assessments. Ask "what's the worst that could happen if this system fails or is misused?" and "what scenarios could lead to that?" For high-impact AI, conduct stress tests and failure mode analysis.

For autonomous vehicles, simulate sensor failures, extreme weather, or contradictory inputs. In finance, regulators require scenario analysis for algorithmic trading under extreme market conditions. By identifying potential failure scenarios, you can design safeties and controls, much like airlines have redundant systems for critical functions.

Maintain AI risk registers with clear ownership. Similar to enterprise project risk registers, maintain an AI risk register documenting identified risks for each AI system, their likelihood, potential impact, and mitigation measures. Keep this updated as systems evolve and assign owners to each risk. This ensures no serious risk remains "out of sight, out of mind." LinkedIn's governance recommendations explicitly mention documenting potential AI risks and mitigation plans as key practice. Link risk registers to your AI inventory, creating comprehensive risk visibility.

Develop incident response playbooks for speed. Incident response playbooks for AI-related issues outline step-by-step actions for plausible failure scenarios. If AI managing your power grid detects anomalies or malfunctions, what immediate actions should engineers take? Who has authority to shut down the AI or switch to manual mode? For public-facing crises (e.g., biased results trending on social media) who manages communications? What's the holding statement? Prepare these in advance, including template communications, technical rollback procedures, and responsible personnel. Playbooks should specify clear roles: technical teams, PR/communications, legal, HR (for internal issues), with defined escalation chains. Many companies integrate AI incidents into broader disaster recovery or business continuity plans, though AI's unique aspects warrant dedicated attention.

Implement robust monitoring and alert systems. Deploy comprehensive monitoring on AI outputs and performance. If chatbots experience spikes in negative user sentiment or behavioral deviations, alerts should notify human supervisors. When prediction models' error rates exceed thresholds, trigger immediate investigation. For operational AI (i.e., manufacturing, utilities), implement automated fail-safes: if AI controlling processes gives anomalous commands, systems default to safe modes or request human confirmation. Many accidents are preceded by warning signs such as unusual outputs or dropping model confidence. Real-time reporting of AI system confidence and key metrics helps catch malfunctions early. Designate internal "AI watch teams" to monitor mission-critical AI systems, analogous to site reliability engineering (SRE) teams for digital services.

Conduct incident drills and simulations regularly. Treat AI incident drills with the same rigor you apply to cyber breach simulations. At a global financial services firm, quarterly AI red-team exercises simulate adversaries subtly manipulating a trading algorithm to bias orders. The drill runs on a 24-hour clock: security flags the anomaly, product and data science validate model integrity, technology leaders decide whether to invoke the model kill switch and roll back to a safe baseline, legal assesses regulatory disclosure requirements, and communications prepares client and market statements. Executives chair a rapid decision forum to confirm escalation thresholds, decision rights, and cross-functional handoffs.

These rehearsals expose gaps in playbooks and controls specific to AI: missing model integrity checks, unclear rollback paths, weak prompt/data provenance, and ambiguous internal/external messaging. As with breach tabletops a decade ago, AI incident drills are fast becoming hygiene for operators at scale. Run them regularly, rotate scenarios (bias accusations going viral, output poisoning, prompt-injection on customer channels), and refine the playbook after each exercise to build true organizational muscle memory.

Establish clear kill-switch procedures. For critical systems, implement "kill-switch" or fallback modes that can override or shut off AI if behaving erratically or suspected of compromise. Trading firms must be able to disconnect AI algorithms from markets quickly during rogue trading. Self-driving cars have multiple redundancy layers returning control to humans or simpler safety systems upon main AI failure. Define clearly who can invoke kill-switches and under what conditions. Crisis often deepens when no one is sure who has authority to halt systems. Also maintain manual backup plans; if AI is down, can humans or simpler systems perform essential functions temporarily? Complete manual reversion might be impossible in AI-first worlds, but even degraded operation beats total failure.

Communicate transparently during crises. When AI crises affect customers or public, transparent and empathetic communication preserves trust. If AI errors cause harm or inconvenience, apologize and explain what's known without excessive defensiveness. Emphasize human accountability: "We at Company X take responsibility for the error caused by our AI system, and we are working to fix it immediately." Blaming algorithms appears evasive; stakeholders expect proper oversight. Outline immediate steps (e.g., system shutdown, fix in progress) and longer-term prevention measures. Early honesty prevents reputational damage from compounding. For data breaches or bias incidents, be upfront then focus on remedies. As AI incidents become more common, public scrutiny intensifies regarding responsible handling. Companies known for addressing issues transparently gain competitive differentiation in trust.

Conduct root cause analysis for systematic learning. After managing crises, thorough post-mortem or root cause analysis (RCA) identifies not just technical failures but process or oversight gaps. Was there a testing blind spot? Did teams override governance controls under pressure? Were warning signs missed? Share learnings organization-wide without blame. Some companies contribute to "AI incident databases" internally or externally for industry learning. The Partnership on AI maintains a public AI Incident Database where companies and researchers contribute failure accounts helping others learn. Update guidelines and training incorporating lessons ensures mistakes aren't repeated. As risk managers say, "Every small incident is a dress rehearsal for a larger one."

Consider insurance and regulatory engagement. As risk mitigation evolves, explore AI-related insurance coverage where available. Cyber insurance might cover certain AI cyber incidents. Errors & omissions insurance could cover AI error liabilities under some conditions. The insurance industry continues adapting to AI risks by engaging with insurers provides insights since they often require risk audits. In high-stakes fields, maintain open communication with regulators. Proactive transparency and collaboration during incidents builds trust and may lead to more understanding responses. As oversight increases (EU AI Act will require incident reporting for certain

AI), having internal processes ready for compliance prevents additional trouble.

Consider a hypothetical scenario: A large e-commerce company's AI pricing algorithm inadvertently colludes with a competitor's AI, causing prices to surge dramatically. Social media erupts accusing companies of price gouging, and regulators announce antitrust probes.

Multi-front crisis response would include:

- Immediately disable or adjust the AI pricing algorithm
- Issue statement explaining unintended algorithmic interaction while pledging regulatory cooperation
- Offer refunds to affected customers
- Urgently reconfigure pricing model avoiding competitor price reactions
- Strengthen oversight on algorithmic pricing strategies across departments

Swift, responsible action transforms damaging incidents into examples of accountability and improvement.

Crisis management for AI proves fundamentally interdisciplinary. It requires technical teams, legal, PR, compliance, and critically, leadership decision-making under pressure. Pre-forming "AI incident tiger team" rosters (of cross-functional groups empowered for rapid action) saves crucial time during actual crises.

Organizations handling AI crises well can actually strengthen trust, demonstrating competence and responsibility. Those handling poorly risk immediate fallout plus chilling effects on further AI innovation as executives become AI-wary. Having navigated these operational challenges, we turn finally to how leadership itself must evolve for the AI-first future.

LEAD THE AI-FIRST FUTURE

Tomorrow's successful executives blend technological fluency with deeply human skills.

As a World Economic Forum article succinctly put it, "Just as every business has become a technology business, every executive must now become a technology executive." This doesn't mean every CEO needs to code or every CMO needs a PhD in AI, but they must understand AI's potential and limitations well enough to make informed strategic decisions.

Leaders should be conversant in topics like data quality, model bias, and AI capabilities relevant to their industry. In practice, this might involve formal education, reverse mentoring where young tech-savvy employees mentor senior leaders on digital trends, or bringing AI experts into strategy discussions regularly. Leaders who underestimate AI's impact risk poor decisions or missed opportunities. Conversely, those who grasp implications can integrate AI into vision and strategy naturally.

Soft skills become even more critical. Leaders must ensure organizations use AI aligned with values and serving stakeholders. As AI accelerates work pace and change, leaders serve as stabilizing forces by listening to employee concerns, articulating purpose beyond technology, and ensuring "human touch" remains in customer experiences.

Workers expressing high trust in senior leaders are far more likely to welcome AI advances. Leadership in an AI era involves augmentation, not abdication: using AI insights for better decisions while exercising human judgment where it counts. An AI might flag customer data trends, but leaders weigh that against context like recent events and brand considerations, deciding courses of action AI alone wouldn't dictate.

Strategic roles evolve. Leaders increasingly frame business strategies around AI capabilities. This means identifying where AI creates competitive advantage (such as customer personalization, operational efficiency, innovation speed) and allocating resources accordingly. Some organizations adopt "AI-first strategy," designing products and services with AI at the core rather than add-on. Leaders must balance innovation versus risk: pushing envelopes leveraging AI for growth while instituting guardrails avoiding ethical or compliance pitfalls.

Leadership teams and C-suite composition. New roles like Chief Data Officer (CDO) and Chief AI Officer (CAIO) are emerging. In some, CIO or CTO roles expand covering AI strategy. Deloitte predicts equipping C-suites with AI expertise is essential for driving transformation. Having AI or data experts on boards may soon become as common as financial experts on audit committees. Leadership teams might include "AI councils": senior leaders focused on AI governance and innovation, bridging technical teams and executive committees. The exact structures vary, but silos between "business leadership" and "technical leadership" must dissolve. AI strategy is business strategy, necessitating domain merging. Companies where CEOs and executive teams converse fluently about AI outpace those relegating it to IT departments.

Decision-making processes transform. Leaders access far more sophisticated analytics and predictive insights through AI. This enables more data-driven decision-making at top levels. Boards might review AI-generated risk analyses or scenario simulations during deliberations. Some organizations employ AI tools in strategy planning, using AI to model market responses to different product launch strategies. However, leaders must guard against over-reliance on AI. A phenomenon called automation bias occurs when people defer excessively to algorithmic suggestions. Strong leaders question and validate AI insights, asking teams how AI arrived at answers and encouraging explainable AI use for critical decisions.

The "centaur model" from chess, where human-AI teams outperform either alone, provides a useful framework. Apply AI for processing huge data and pattern finding, humans for judgment in novel situations, ethical considerations, and creativity. Future CEOs might regularly consult AI advisors providing briefings on resource allocation given dozens of variables, but also consult human advisors and their own intuition before finalizing decisions.

Cultural transformation requires active leadership. In AI-first organizations, culture should encourage experimentation and agility. Leaders foster environments where employees aren't afraid trying new AI tools, where failure in experimentation becomes learning rather than something hidden.

Eighty percent of C-suite executives globally believe AI will spark culture shifts making teams more innovative. But shifts don't happen automatically; leaders catalyze by modeling behavior. When COOs share how they used AI-driven dashboards for decisions and praise teams building them, it signals using AI creatively is valued. "Lead from the front" applies: when top executives personally engage with AI and champion use, it cascades organization-wide.

Ethical and responsible leadership. With AI's power comes heightened scrutiny. Public and employees expect leaders ensuring AI is used ethically and beneficially. Future leaders will be judged not just on financial metrics but on how responsibly they deploy AI: avoiding misuse, respecting privacy, mitigating bias, considering societal impacts. Being proactive becomes part of visionary leadership, akin to how environmental sustainability grew from side concern to core leadership agenda. Leaders must set the tone that capability doesn't equal obligation; just because we can do something with AI doesn't mean we should.

Establish clear ethical guidelines and take a stand on AI-related public policy. Examples include tech leaders deciding against certain surveillance AI contracts for ethical reasons, or lobbying for thoughtful regulation ensuring level playing fields and public trust.

CEOs in AI-first eras might be as concerned with their company's AI "social license to operate" as quarterly sales.

Continuous learning. With AI's rapid evolution, leaders must learn continuously through formal means or informal approaches like reading, attending AI conferences, or having technology advisors. Effective AI-era leaders are "learn-it-alls" rather than "know-it-alls". They encourage curiosity in themselves and others. Nadella transformed Microsoft's culture emphasizing learning over static expertise, an approach well-suited for ongoing AI advances. Practically, leadership teams might allocate meeting time discussing emerging tech trends and strategic implications. They might set personal learning goals like "this year I will grasp neural network fundamentals" or "I will mentor with data scientists quarterly."

Leadership styles shift toward collaboration and decentralization. With AI empowering more autonomous decision-making at lower levels through analytics, command-and-control leadership fades further. Leaders create visions and guidelines, empowering teams to execute often aided by AI tools providing enhanced insight. Leaders' roles become more about orchestration and inspiration than micromanagement. AI can flatten knowledge hierarchies; sometimes frontline employees armed with AI insights spot opportunities higher-ups miss. Smart leaders listen and harness collective intelligence. We see emergence of data-driven democratic styles where leaders rely on transparent dashboards and input from all levels steering organizations.

Resilience and adaptability. New competitors arise quickly leveraging AI, market conditions shift with algorithmic influences, global events change data patterns overnight disrupting AI models. Leaders must navigate volatility maintaining composure during AI-related crises. Incorporate scenario planning including AI disruptions, preparing strategic responses. Adaptability applies to leaders' own roles: as AI takes routine analysis, leaders focus more on areas AI can't address: motivating people, negotiating complex agreements, envisioning unique products.

Eighty-eight percent of global executives say helping businesses speed up AI adoption is an important priority, showing most top leaders realize their championing role. About 80 percent believe AI will unlock more innovative team cultures, suggesting optimism that AI can make organizations not just efficient but creative and dynamic under right leadership.

Consider a future leadership scenario: A CEO in 2030 starts her day with an AI brief generated overnight, summarizing key performance metrics, flagging anomalies, prioritizing insights by estimated importance. During executive meetings, the Chief AI Officer presents AI-simulated outcomes of proposed strategy shifts. Leaders discuss findings, question assumptions, then integrate them with market intuition.

The HR head reminds teams of ethical guidelines when proposed marketing AI approaches privacy boundaries; the CEO and team adjust approaches aligning with principles. Later, during employee town halls, she shares stories of teams combining AI insights with human creativity to solve customer problems, highlighting that synergy as company's future. Fielding questions about AI job impacts, she responds candidly about reskilling commitments and new role creation. That evening, she completes online AI strategy course modules, part of ongoing education commitment.

This portrait shows leaders comfortable using AI as tools, deeply aware of human impacts, steering culture and strategy in tandem with technology. The essence of leadership (providing vision, inspiration, and direction) remains unchanged, but the context transforms. In the future, leadership will be defined less by what you know (since AI can know so much) and more by how you think and how you behave.

Interpersonal aspects such as motivating diverse AI-empowered workforces, building trust with customers in AI-driven services, and collaborating across boundaries become true leadership art. We see positive feedback loops emerging: strong leadership enables better AI adoption, and effective AI use strengthens organizational performance, validating leadership vision.

Leaders who excel in this environment will shape not just their organizations but the broader AI revolution itself. They'll demonstrate that properly led, AI amplifies rather than replaces human potential.

SUMMARY

This chapter outlined the advanced operating system that separates AI leaders from those stuck in pilot purgatory. The seven interconnected domains (regulation, governance, ethics, workforce, sustainability, risk management, and leadership) operate as a single system, not a checklist, to scale AI with speed and safety.

Reaching the success tier requires maturity across all seven. Companies that treat EU AI Act compliance as strategic infrastructure create defensible moats. Those that implement comprehensive governance (anchored in NIST frameworks) and empower AI ethics boards innovate faster with guardrails. Organizations that operationalize ethics with systematic bias testing, human-in-the-loop oversight, and transparent documentation convert trust into an enduring asset.

Workforce transformation is equally decisive: with potentially 40 percent of workers requiring reskilling and 70 percent of job skills shifting by 2030, firms that build AI academies and stand up new roles turn displacement risk into competitive advantage. To sustain momentum, leaders pair relentless ROI proof points with durable MLOps foundations and ongoing executive sponsorship, shifting AI from initiative to operating model.

Chapter takeaways:

- Scaling AI requires moving decisively from isolated pilots to enterprise-wide adoption, integrating solutions into core workflows and strategic priorities.
- Investment must rebalance over time, shifting emphasis from technology acquisition to people, process optimization, and governance maturity.
- Cross-functional collaboration becomes non-negotiable, breaking down silos and embedding shared accountability for outcomes.
- Change management is treated as a cultural imperative, with structured communication and active stakeholder engagement driving adoption.
- Success metrics evolve beyond technical performance to business impact, ensuring executive sponsorship remains anchored in measurable value.

Understanding these frameworks is only the beginning. The real challenge lies in translating vision into action: bridging the persistent gap between AI strategy and execution. In the next chapter, we shift from theory to practice, examining how organizations can operationalize their AI ambitions, align initiatives with business objectives, and build the execution muscle required to move from pilot purgatory to enterprise-wide impact.

9

IMPLEMENTING AI TRANSFORMATION – FROM STRATEGY TO EXECUTION

According to publicly available reports at the time this book was written, only a small fraction of companies were realizing significant returns from their AI investments, while most witnessed promising pilot projects fail to scale. The primary reason cited in these reports is a widespread misconception: many organizations approached AI transformation as a technology initiative rather than a holistic organizational evolution.

CEOs and boards face a critical inflection point. Those who master AI execution will dominate their industries. Those who stumble join the graveyard of digital transformation failures: Polaroid, Blockbuster, and countless others who confused motion with progress. As one BCG study revealed, organizations that successfully operationalize AI are five times more likely to see substantial business impact than those stuck in perpetual experimentation.

The road from AI vision to value surgical precision in aligning initiatives with strategy, architecting implementation roadmaps that anticipate reality's messiness, and building measurement systems that capture actual business impact rather than vanity metrics. Most critically, it demands leadership that understands the difference between deploying AI and transforming through AI.

This chapter cuts through the complexity with a practical framework for AI execution. You'll learn how to forge unbreakable links between AI projects and business objectives, construct roadmaps that survive first contact with organizational reality, and establish metrics that matter to shareholders. We'll explore resource allocation that maximizes ROI, scaling strategies that break free from pilot purgatory, and the continuous learning systems that separate AI leaders from laggards.

By chapter's end, you'll possess the blueprint for transforming AI from expensive experiment to strategic differentiator, ensuring your organization shapes the AI era rather than becoming its casualty.

ALIGNING AI INITIATIVES WITH BUSINESS STRATEGY

The foundation of every successful AI transformation rests on one principle: technology serves strategy, not the reverse. Yet most organizations begin their AI journey by asking what the technology can do, rather than what the business needs. This backwards approach explains why scattered AI experiments proliferate while strategic impact remains elusive.

AI becomes a strategic weapon only when aimed at your organization's most critical objectives. Consider how a major bank transformed its approach: instead of launching dozens of disconnected chatbot projects, leadership identified two core strategic priorities: improving customer retention and reducing fraud losses. Every AI investment had to demonstrably advance these goals. Within 18 months, AI-driven personalization increased customer lifetime value by 23 percent, while machine learning models cut fraud losses by $47 million annually.

The alignment process begins with brutal clarity about your strategic North Star. What drives competitive advantage in your industry? Where do operational inefficiencies threaten profitability? Which customer pain points offer the greatest opportunity? These questions must precede any discussion of algorithms or platforms. As one Fortune 500 CEO observed, successful AI initiatives start in the boardroom, not the data science lab.

When C-suite executives view AI as someone else's project (i.e., IT's), initiatives drift toward technical showcases rather than business solutions. Establish AI governance at the highest levels, with direct CEO /CAIO involvement and clear accountability cascading through business units. Titles matter less than ensuring AI strategy flows from business strategy, not parallel to it.

Creating this alignment requires answering three deceptively simple questions before approving any AI project:

- Does this initiative directly support our stated strategic priorities?
- Can we articulate the specific business outcome we expect?
- Have we engaged the business stakeholders who will own the results?

Projects that can't provide compelling answers deserve scrutiny, regardless of their technical elegance. A global retailer learned this lesson expensively: their AI-powered robotic greeter impressed visitors but contributed nothing to strategic goals of supply chain optimization and customer loyalty. When leadership enforced strategic alignment, resources shifted to AI systems that reduced inventory costs by 15 percent and increased repeat purchases through personalized recommendations.

Business unit leaders naturally resist AI initiatives they perceive as imposed from outside. But when AI projects emerge from their own strategic challenges with their teams involved from conception, adoption accelerates dramatically. Smart organizations create "pull" for AI by helping business leaders identify where intelligent automation could unlock their specific goals, rather than pushing generic AI solutions.

This alignment discipline extends beyond project selection to ongoing governance. Regular reviews should examine whether AI initiatives maintain their strategic relevance as business conditions evolve. Markets shift, competitors move, and yesterday's strategic priority may become today's distraction. The ability to pivot or terminate AI projects that lose strategic alignment often distinguishes organizations that generate ROI from those that accumulate technical debt.

With strategic alignment established, the temptation to sprint toward implementation becomes overwhelming. Yet organizations that yield to this impulse consistently stumble. The path from AI vision to operational reality demands careful choreography: a roadmap that anticipates obstacles, sequences dependencies, and maintains momentum through inevitable setbacks.

CREATING A COMPREHENSIVE AI IMPLEMENTATION ROADMAP

The most effective AI roadmaps resemble military campaign plans more than technology project schedules. They account for terrain (organizational readiness), supply lines (data and talent), and potential resistance (change management). Most critically, they phase the journey into winnable battles rather than attempting wholesale transformation.

Consider how a global manufacturer approached their AI transformation. Rather than declaring a sweeping "Industry 4.0" initiative, they mapped a three-phase journey. Phase One focused on foundational elements: data infrastructure consolidation, pilot projects in predictive maintenance, and building an AI Center of Excellence. Phase Two expanded successful pilots to multiple facilities while adding new use cases in quality control and supply chain optimization. Phase Three integrated AI capabilities into core business processes and customer-facing products. Each phase had clear success criteria and decision gates before proceeding.

Effective roadmaps break down into several critical components:

Phases and Milestones: Structure your journey into digestible segments, typically 3-6 months each. Early phases emphasize foundation-building and quick wins that build credibility. Middle phases tackle more complex integrations and scaling challenges. Later phases focus on optimization and innovation. Each phase should conclude with measurable milestones that trigger go/no-go decisions for the next stage.

Data and Technology Foundation: AI systems consume data voraciously and demand robust infrastructure. Your roadmap must sequence data preparation activities like cleaning, integration, and governance ahead of algorithm development. A healthcare system learned this painfully when their AI diagnostic project stalled for eight months while teams retroactively standardized patient data formats across disparate systems.

Skills and Team Building: The scarcest resource in AI transformation isn't technology; it's talent. Roadmaps must account for hiring timelines, training programs, and team formation. Leading organizations adopt a hybrid approach: recruiting external AI specialists while simultaneously upskilling internal staff who possess invaluable domain knowledge. One bank's roadmap included deliberate knowledge transfer periods where external consultants worked alongside internal teams, gradually shifting ownership.

Governance Checkpoints: Every AI system introduces new risks, from algorithmic bias to security vulnerabilities. Build governance reviews into your roadmap at natural transition points. These are essential moments to assess whether your AI initiatives remain aligned with strategic goals, ethical principles, and regulatory requirements. A pharmaceutical company's roadmap included formal ethics reviews before any AI system touched patient data or influenced treatment recommendations.

Change Management: Technical implementation represents perhaps a third of the transformation challenge. The remaining two thirds involves human adaptation. Successful roadmaps explicitly map change management activities: communication campaigns, training programs, pilot user groups, and feedback mechanisms. They identify potential resistance points and plan interventions accordingly.

The roadmap must remain a living document, not a rigid scripture. Market conditions shift, technologies evolve, and organizational priorities change. Build in quarterly reviews where leadership can adjust timelines, reprioritize initiatives, or pivot approaches based on lessons learned. This adaptability proved crucial for a retailer whose original roadmap emphasized in-store AI applications before COVID-19 forced a rapid pivot toward e-commerce and supply chain applications.

Resource allocation deserves special attention in roadmap design. Avoid the common mistake of front-loading all investment in year one. Instead, structure funding to follow proven success, with increasing investment as initiatives demonstrate value. This approach maintains CFO support while providing flexibility to redirect resources from underperforming projects.

Global enterprises must decide whether to pursue uniform worldwide deployment or allow regional variation. Similarly, highly diversified companies face choices about centralized versus distributed AI capabilities. The most successful approaches typically combine standardized platforms and governance with localized implementation and use cases.

Identify potential failure points such as key person dependencies, technology lock-in, and regulatory changes, and design contingencies. One industrial company's roadmap included parallel workstreams for critical initiatives, ensuring that delays in one area wouldn't cascade across the entire program.

The roadmap provides the journey's route, but without clear metrics, organizations navigate blind. Too many AI initiatives drown in technical metrics that impress data scientists but leave executives cold. The disconnect between algorithmic performance and business impact explains why boards lose patience with AI investments that show theoretical promise but deliver minimal bottom-line results.

ESTABLISHING CLEAR METRICS AND KPIS FOR AI SUCCESS

Meaningful AI metrics bridge the chasm between technical achievement and business value. They translate model accuracy into customer satisfaction improvements, processing speed into operational cost savings, and prediction precision into revenue growth. This translation requires thoughtful design and continuous calibration.

A telecommunications company learned this distinction painfully. Their AI customer service system boasted 95 percent accuracy in understanding customer queries. Yet customer satisfaction scores barely moved. Deeper investigation revealed the AI correctly understood questions but provided responses customers found frustrating or incomplete. Only when they shifted metrics to "first-call resolution rate" and "customer effort score" did meaningful improvements emerge.

Effective AI metrics operate at three distinct levels:

Business Outcome Metrics: These connect directly to strategic objectives and financial performance. For a credit card company using AI for fraud detection, business metrics might include fraud losses prevented, false positive rates affecting customer experience, and operational cost savings from reduced manual reviews. These metrics resonate in boardrooms and earnings calls.

Operational Performance Metrics: These measure AI's impact on business processes. An AI-powered supply chain optimization system might track inventory turns, stockout incidents, and order fulfillment accuracy. While more granular than business outcomes, these metrics maintain clear links to value creation.

Technical Performance Metrics: These assess the AI system's functional capabilities. Model accuracy, precision, recall, and processing latency matter for system health but shouldn't dominate executive dashboards. Technical metrics serve as early warning systems for degradation but require translation into business language for broader consumption.

The art lies in connecting these three levels. A retail bank created a measurement cascade for their AI loan approval system:

Technical metric: Model prediction accuracy (target: 92 percent)

Operational metric: Loan processing time (target: 4-hour reduction)

Business metric: Revenue from increased loan volume (target: $15M annually)

When model accuracy unexpectedly dropped to 89 percent, they could immediately calculate the downstream impact on processing time and revenue, justifying immediate remediation investment.

Before deploying AI, meticulously document current performance levels. How long do processes take today? What error rates do humans produce? What customer satisfaction scores exist? Without baselines, AI's incremental value remains invisible. One manufacturer discovered their AI quality control system performed only marginally better than experienced human inspectors, that is until they measured the system's consistency across shifts, fatigue-free operation, and ability to document every decision.

Early implementations might focus on technical stability and user adoption. As systems stabilize, metrics shift toward business impact and optimization opportunities. Mature deployments emphasize innovation metrics: such as new use cases enabled, adjacent problems solved, or competitive advantages created.

Organizations with strong data cultures can handle nuanced measurement frameworks. Others need simpler, more intuitive metrics that build confidence gradually. A healthcare provider initially resisted AI diagnostic tools until metrics shifted from "algorithm confidence scores" to "physician time saved" and "patient outcomes improved", measures that resonated with medical professionals' values.

Any single metric, pursued too aggressively, creates perverse incentives. Customer service AIs optimized solely for call duration might rudely rush customers off the phone. Fraud detection systems focused exclusively on catching fraudsters might flag legitimate transactions excessively. Balanced scorecards that include competing metrics (e.g., efficiency and satisfaction, precision and recall) prevent optimization disasters.

Real-time dashboards seduce with their immediacy but can trigger overreaction to normal variations. Monthly or quarterly reviews often provide better signal-to-noise ratios for strategic metrics, while operational metrics might warrant weekly examination. Technical metrics typically require continuous monitoring with automated alerting for anomalies.

Document exactly how each metric derives from raw data. When metrics require adjustment, explain changes thoroughly. One financial services firm publishes a "metrics methodology" document updated quarterly, ensuring all stakeholders understand what numbers mean and how they're calculated.

Integration with existing performance management systems accelerates adoption. Rather than creating standalone AI scorecards, weave

AI metrics into established business reviews, operational dashboards, and performance evaluations. This integration reinforces AI's role as business enabler rather than technical curiosity.

The metrics foundation sets the stage for one of AI transformation's most challenging aspects: securing and allocating the resources needed to turn plans into reality.

MANAGING RESOURCES AND BUDGETING FOR AI INITIATIVES

Many Global 2000 enterprises now pour over $10 million annually into AI efforts, yet investment alone guarantees nothing. The difference between value creation and expensive experimentation lies in how resources are allocated, managed, and adapted.

Smart AI resource management resembles venture capital more than traditional IT budgeting. Portfolio thinking replaces project-by-project evaluation. Organizations that demand positive ROI from every AI experiment paradoxically achieve lower overall returns than those accepting intelligent failure as the price of breakthrough innovation.

A European bank's transformation illustrates this principle. Rather than spreading AI investment evenly across divisions, they established a central AI fund with clear allocation criteria: 40 percent for proven use cases ready to scale, 40 percent for promising pilots with strong business sponsorship, and 20 percent for moonshot experiments that could reshape banking. This portfolio approach balanced risk while ensuring resources flowed to highest-impact opportunities.

Resource planning encompasses multiple dimensions:

Financial Architecture: Traditional annual budgeting cycles clash with AI's iterative nature. Leading organizations adopt rolling funding models with quarterly gates. Initial investments need only fund proof-of-concept development and limited pilots. Success unlocks progressively larger tranches. This staged approach maintains CFO confidence while providing flexibility to pivot or accelerate based on results.

Infrastructure Investment: The temptation to delay infrastructure spending until AI proves itself creates a vicious cycle: underpowered systems produce underwhelming results, justifying continued underinvestment. Smart organizations front-load foundational investments in data platforms, cloud computing capabilities, and development tools. One retailer's $15 million data lake investment seemed excessive until it enabled dozens of AI use cases, each impossible under the previous fragmented architecture.

Talent Economics: Human capital often represents AI's largest expense, and greatest constraint. Beyond data scientists' eye-watering salaries, consider the full talent ecosystem: ML engineers for deployment, data engineers for pipeline construction, domain experts for context, and change managers for adoption. Use creative approaches help manage costs such as rotational programs that expose existing employees to AI projects, partnerships with universities for research collaboration, and strategic use of external expertise for capability building rather than permanent staffing.

Organizational Structure Implications: Resource allocation decisions shape organizational design. Centralized AI teams achieve economies of scale and consistent standards but risk disconnect from business realities. Fully distributed models embed AI deeply in business units but sacrifice coordination and knowledge sharing. Consider hybrid models like centers of excellence providing platforms and governance while business units own specific applications.

Budget scenarios should reflect AI's inherent uncertainty. Build multiple funding pathways based on early results. If pilot projects exceed expectations, how quickly can additional resources deploy? If initiatives stall, what's the minimum investment to preserve optionality? One pharmaceutical company created three budget scenarios with predefined triggers for shifting between them based on pilot outcomes and competitive moves.

Data preparation consumes 60–80 percent of AI project timelines due to the intensive work of cleaning, labeling, and integrating data. Integration with legacy systems adds further complexity; BCG notes that many enterprises still rely on decades-old software, making modernization efforts costly and time-consuming. Meanwhile, successful AI adoption hinges on sustained investment in change management and workforce training, not just one-off rollouts. Organizations that budget only for visible algorithm development often underestimate the true cost of AI; the most successful ones explicitly plan for these "underwater iceberg" expenses to ensure long-term scalability and impact.

High-risk initiatives (i.e., those involving sensitive data, customer-facing applications, or regulatory complexity) require contingency reserves. One bank allocated 25 percent budget premiums for AI projects touching customer data, funding additional privacy protections and compliance reviews. This proactive approach prevented costly retrofitting when regulations tightened.

ROI measurement requires nuanced thinking. Pure financial returns tell only part of the story. Strategic benefits (e.g., improved customer experience, enhanced decision-making speed, new capability development) deserve consideration even when difficult to quantify precisely. Develop "balanced scorecards" for AI investments, weighing financial metrics alongside strategic impact indicators.

Vendor management presents unique challenges in AI procurement. The landscape shifts rapidly; yesterday's cutting-edge platform becomes tomorrow's legacy technology. Avoid long-term lock-in through modular architectures and portable standards. Negotiate contracts that anticipate change: scaling provisions, technology refresh rights, and clear exit strategies. One manufacturer's prescient contract terms allowed seamless migration when their original AI vendor was acquired, avoiding months of disruption.

International operations complicate AI resource planning due to regional differences in talent costs, data privacy regulations, and infrastructure maturity. Global organizations must weigh the benefits of centralized centers of excellence against the agility of distributing capabilities closer to local markets. Strategic decisions are further shaped by currency volatility, tax structures, and intellectual property protections, all of which can significantly impact cost, compliance, and innovation velocity.

Budget governance requires delicate balance. Too much oversight stifles innovation; too little invites waste. Establish clear decision rights: who can approve pilots, who authorizes scaling, who decides on platform investments. Regular portfolio reviews assess overall AI investment health rather than micromanaging individual projects. One technology company's quarterly "AI investment board" brings together finance, technology, and business leaders to evaluate portfolio performance and adjust allocations.

Securing resources marks a crucial milestone, but the journey's next phase presents an equally formidable challenge: scaling successful pilots across the enterprise without losing the magic that made them work.

SCALING AI PILOTS ACROSS THE ORGANIZATION

The graveyard of AI initiatives overflows with brilliant pilots that never escaped the lab. A chatbot that delighted test users becomes a customer service nightmare at scale. A fraud detection algorithm that saved millions in one region creates false positive chaos when deployed globally. The leap from pilot success to enterprise impact breaks more AI dreams than any technical challenge.

Scaling AI resembles organ transplantation more than software deployment. Success requires matching donor and recipient, suppressing rejection responses, and monitoring constantly for complications. Organizations that treat scaling as simple replication discover that what worked in controlled conditions fails catastrophically in the wild.

A global insurance company's AI underwriting assistant showed remarkable results in the Nordic pilot: processing time decreased 40 percent, accuracy improved 15 percent, and underwriters embraced the technology. Initial scaling attempts failed spectacularly. The UK deployment crashed against different regulatory requirements. The US version couldn't handle state-by-state variations. Asian markets rejected the interface design. Only when leadership shifted from "copy-paste" to "adapt and evolve" (tailoring deployments to local conditions) did successful scaling emerge.

Effective scaling strategies address multiple dimensions simultaneously:

Technical Architecture for Scale: Pilot systems often run on jury-rigged infrastructure adequate for hundreds of users but catastrophic for millions. Scaling demands industrial-strength architecture: cloud-native design, microservices for modularity, API-first integration, and robust monitoring.

Data Pipeline Evolution: Pilots typically operate on carefully curated datasets. Production systems must handle messy, incomplete, rapidly changing data streams. A healthcare AI that diagnosed rare diseases from pristine research images failed when fed the varied quality scans from hundreds of hospitals. Successful scaling requires "data hardening", or building systems resilient to real-world data's imperfections.

Local Adaptation Without Fragmentation: The paradox of scaling lies in maintaining consistency while accommodating variation. Different markets, regulations, and business contexts demand flexibility. Yet allowing unlimited customization creates an unmaintainable mess. Leading organizations establish "core and flex" architectures: standardized platforms with well-defined customization points. A multinational bank's credit scoring AI maintains consistent risk models globally while allowing regional adjustments for local economic conditions.

Change Management at Scale: Pilot users are typically enthusiasts who volunteer for innovation. Scaled deployment faces skeptics, resisters, and the indifferent. Communication strategies must segment audiences: executives need business impact stories, middle managers require integration roadmaps, and front-line users want practical training. Success hinges on meeting each group where they are by addressing their specific concerns, incentives, and workflows.

Governance Evolution: Pilot governance often relies on informal oversight and rapid decision-making. Scale demands formal structures: clear ownership models, escalation pathways, and risk management protocols. A telecommunications company established an "AI Operations Center" as they scaled their network optimization system: a dedicated team monitoring performance, managing incidents, and coordinating updates across regions.

Performance Monitoring Complexity: Pilots generate manageable data volumes for performance tracking. Scaled systems produce overwhelming metric floods. Effective monitoring systems must separate signal from noise through intelligent aggregation, anomaly detection, and automated alerting. Manual review becomes impossible at scale; automated systems must flag deviations requiring human attention.

Incremental Rollout Strategies: Phased approaches by geography, business unit, or customer segment allow controlled expansion with learning opportunities between phases. A payment processor scaled their fraud detection AI through careful staging: starting with low-risk merchant categories, expanding to medium-risk segments after proving stability, and finally tackling high-risk transactions with lessons learned from earlier phases.

Knowledge Transfer Mechanisms: Successful organizations create systematic knowledge transfer: documentation, training programs, and rotation assignments spreading know-how. One pharmaceutical company required pilot team members to spend 25 percent of their time supporting scaling efforts, ensuring lessons learned propagated throughout the organization.

Economic Model Refinement: Per-unit costs are expected to decrease with scale, but hidden expenses often emerge, ranging from increased support needs and integration complexity to compliance overhead. Financial models must evolve from optimistic pilot projections to reflect production realities. For example, a logistics company implemented an AI-powered route optimization system that delivered impressive cost savings in urban test markets. However, when scaled to rural regions, the model underperformed due to sparse delivery networks and inconsistent data coverage. Only after retraining the algorithm with rural-specific constraints and adjusting the cost model to account for longer travel distances and lower delivery density did the deployment become financially viable.

Cultural Bridge-Building: Different organizational cultures embrace AI differently. Engineering-heavy divisions adopt readily; traditional operations resist change. Successful scaling requires cultural translation, or framing AI benefits in terms each audience values. A utility company positioned the same predictive maintenance AI as "innovation enabler" for technology teams and "risk reducer" for operations groups, tailoring messages to cultural priorities.

Platform Thinking: Organizations running multiple AI initiatives often struggle when each project uses its own tools and infrastructure. Instead of relying on isolated, one-off solutions (point solutions), they benefit more from a platform approach (a shared foundation of tools and services that can be reused across projects). These platforms typically include common components like data pipelines, model training environments, and deployment frameworks. By standardizing these elements, teams avoid reinventing the wheel each time and can scale new use cases much faster. For example, one retailer that invested in an enterprise-wide AI platform reduced the time to launch new AI use cases from six months to just six weeks, thanks to reusable infrastructure and streamlined processes.

Ecosystem Considerations: Scaled AI systems exist within complex organizational ecosystems. Integration with existing systems, data sources, and business processes determines success more than algorithmic sophistication. A manufacturing AI that optimized individual plant operations actually reduced overall supply chain efficiency until integration expanded to include upstream suppliers and downstream distribution.

Risk Amplification: Problems acceptable in pilots become crises at scale. A bias affecting dozens of pilot users becomes discrimination impacting thousands. Security vulnerabilities exposing limited test data become breaches compromising millions of records. Scaling must include proportional investment in risk management, security hardening, and ethical safeguards.

Successfully moving from pilot to production is a major milestone, but sustaining AI value at scale requires the capability to continuously learn, adapt, and optimize systems as they operate in dynamic, real-world environments.

CONTINUOUS LEARNING AND FEEDBACK LOOPS FOR AI OPTIMIZATION

AI systems require metabolism: constant intake of new information, processing of feedback, and adaptation to changing conditions. AI systems exist in dynamic environments where customer behaviors shift, market conditions evolve, and data patterns drift. Yesterday's cutting-edge algorithm becomes tomorrow's obsolete relic without systematic mechanisms for continuous improvement. Static AI is dying AI. Organizations that view deployment as the finish line watch their systems decay into expensive technical debt while competitors' solutions grow more capable daily.

A major airline's revenue management system illustrates both the peril and potential. Their AI-driven pricing algorithm, revolutionary when launched, gradually lost effectiveness as passenger behaviors shifted post-pandemic. Competitors gained market share while their system clung to outdated patterns. Only after implementing comprehensive feedback loops (monitoring booking patterns, competitor actions, and demand signals) did performance recover and eventually exceed original benchmarks.

Building effective learning systems requires multiple interconnected components:

Performance Monitoring Architecture: Continuous learning begins with continuous observation. Modern AI systems generate vast telemetry streams: prediction confidence scores, input data characteristics, processing latencies, and outcome accuracy. Effective monitoring systems filter this flood into actionable insights. Automated anomaly detection flags unusual patterns before they cascade into failures. One credit card company's monitoring system detected a subtle drift in fraud patterns three weeks before human analysts noticed, preventing millions in potential losses.

Feedback Integration Mechanisms: The most valuable feedback often comes from system users and affected stakeholders. Yet most organizations lack systematic ways to capture and process this intelligence. One way is to build feedback directly into AI interfaces: simple thumbs up/down ratings, correction mechanisms, and escalation pathways. A healthcare system's diagnostic AI improved dramatically after implementing "physician override" tracking, learning from cases where doctors disagreed with algorithmic recommendations.

Model Retraining Pipelines: Data scientists love building new models but often neglect maintaining existing ones. Production AI requires industrial-strength retraining infrastructure: automated data pipelines feeding fresh training sets, systematic experimentation with model variations, and controlled deployment of updated versions. One retailer automated their recommendation engine retraining, moving from quarterly manual updates to weekly automatic refreshes, increasing relevance and revenue.

A/B Testing Frameworks: The scientific method applies to AI optimization as rigorously as drug development. Systematic experimentation reveals which changes improve performance and which merely add complexity. Sophisticated organizations run hundreds of simultaneous experiments: testing algorithm variants, interface modifications, and process adjustments. A streaming service discovered through careful testing that simpler recommendation algorithms often outperformed complex deep learning models for certain user segments.

Version Control and Rollback Capabilities: AI systems need industrial-strength change management rivaling mission-critical software. Every model version, configuration change, and parameter adjustment requires tracking. When problems emerge, rapid rollback prevents cascading failures. One bank's loan approval AI experienced a sudden accuracy drop; version control allowed instant reversion to the previous stable model while engineers diagnosed the issue offline.

Human-in-the-Loop Learning: Expert knowledge, ethical judgment, and contextual understanding complement machine learning capabilities. Design "human-in-the-loop" systems where expert input directly improves AI performance. A legal AI company's contract analysis system allows senior lawyers to correct misinterpretations; these corrections immediately update the training data, creating a virtuous cycle of improvement.

Cross-Functional Learning Teams: Effective learning systems unite data scientists, domain experts, business analysts, and end users in collaborative improvement efforts. Regular "AI retrospectives" examine performance data, user feedback, and business outcomes to identify optimization opportunities. One manufacturer's predictive maintenance team includes data scientists, reliability engineers, and shop floor technicians, ensuring improvements reflect both technical possibilities and operational realities.

External Signal Integration: AI systems operating in isolation miss crucial environmental changes. Competitive intelligence, regulatory updates, and market trends should inform continuous improvement efforts. An insurance company's pricing AI incorporates external data feeds: competitor rate changes, weather patterns affecting claims, and economic indicators predicting customer behavior. This external awareness prevents the "boiled frog" syndrome where gradual environmental changes go unnoticed.

Ethical and Bias Monitoring: Continuous monitoring is essential for ensuring AI systems remain fair, ethical, and effective as they scale. It must track how performance varies across different demographic groups, detect emerging biases, and flag ethical concerns before they cause harm. These issues often evolve over time as user behavior, language, and data patterns shift, making static evaluations insufficient. Without ongoing oversight, AI systems can unintentionally reinforce inequalities or make flawed decisions that go unnoticed. Proactive monitoring allows organizations to catch these problems early, update models responsibly, and maintain trust with users and stakeholders.

Knowledge Management Systems: Documented lessons learned, pattern libraries, and solution databases accelerate future improvements. A global consulting firm created an "AI Knowledge Hub" where teams worldwide share optimization techniques, failure analyses, and breakthrough innovations, dramatically accelerating the learning curve for new deployments.

Metric Evolution: Success metrics in AI systems must evolve over time to reflect changing priorities and realities. Early deployments often focus on technical benchmarks like accuracy or precision, which are useful for validating initial performance. But as systems mature and scale, organizations must shift toward metrics that capture real-world impact such as user satisfaction, operational efficiency, or business outcomes. This transition requires not only redefining what success looks like, but also building the capability to continuously reassess and refine metrics as user needs, market conditions, and ethical considerations evolve.

Organizational Learning Culture: Technology provides the tools for continuous improvement, but it's organizational culture that determines whether those tools are used effectively. Companies that encourage learning from failure, reward experimentation, and promote open knowledge sharing tend to see far greater progress in their AI initiatives. A culture of curiosity and psychological safety allows teams to surface issues early, iterate quickly, and build on each other's insights. Without this foundation, even the most advanced technologies can stagnate.

Automated Improvement Systems: Advanced techniques like automated hyperparameter optimization, neural architecture search, and meta-learning allow AI systems to continuously explore ways to improve themselves without needing manual adjustments. These methods test different configurations, model structures, and learning strategies in the background, speeding up experimentation and refinement. While they don't replace human oversight, they significantly reduce the time and effort required to enhance performance, making it easier to adapt models to new data or changing conditions. This creates a feedback loop where systems evolve faster and more efficiently, helping organizations stay ahead in dynamic environments.

Competitive Benchmarking: Regular benchmarking is essential to ensure AI systems remain competitive and effective over time. Comparing performance against industry standards, academic advancements, and emerging competitor solutions helps organizations avoid complacency and identify areas for improvement. These benchmarks provide external reference points that challenge internal assumptions and reveal gaps that may not be visible through internal metrics alone. Without this discipline, even well-performing systems risk falling behind as the broader landscape evolves. Continuous benchmarking fosters a mindset of excellence, encouraging teams to refine models, update practices, and stay aligned with the latest innovations and expectations.

Mastering continuous improvement provides sustainable competitive advantage, but even perfectly optimized current systems can't protect against disruption. The final frontier requires preparing for AI capabilities that don't yet exist.

PREPARING FOR THE NEXT WAVE OF AI INNOVATION

Future readiness requires building organizational capacity to absorb and deploy whatever emerges. The half-life of AI dominance shrinks monthly. Today's breakthrough becomes tomorrow's baseline capability as the field accelerates beyond linear progression. Organizations celebrating current AI achievements risk the innovator's dilemma: excelling at yesterday's game while missing the next fundamental shift.

Sustained AI leadership belongs to those who prepare for capabilities not yet invented. The companies that dominated the internet era weren't necessarily the best at building websites; they were the most adaptable as the web evolved from static pages to dynamic platforms to mobile-first experiences. AI's trajectory promises even more dramatic transformations.

Consider how rapidly generative AI shifted from research curiosity to business essential. Organizations that had built flexible AI foundations integrated large language models within months. Those with rigid, narrow implementations watched competitors leapfrog their capabilities overnight. Architectural decisions made today determine adaptive capacity tomorrow.

Building future readiness requires several interlocking strategies:

Technology Horizon Scanning: Establish dedicated teams tracking research breakthroughs, startup activity, and patent filings. Effective scanning requires translating technical advances into business implications. When transformer architectures emerged from Google's research labs, forward-thinking companies immediately began exploring applications, years before ChatGPT made the technology mainstream.

Capability Building Partnerships: No single organization can match the global AI research community's innovation pace. Strategic partnerships with universities, research labs, and startups provide early access to emerging capabilities. Small companies can still achieve benefits through targeted collaborations, joint research projects, and startup incubator programs.

Experimental Budgets: Innovation requires permission to fail. Organizations serious about AI's future allocate specific budgets for exploring unproven technologies. Google's "20 percent time" philosophy applies equally to AI experimentation. One retailer dedicates 15 percent of their AI budget to "future bets": technologies without immediate ROI but potential game-changing impact. This patient capital funds exploration of quantum machine learning, neuromorphic computing, and other frontier technologies.

Talent Pipeline Development: Future AI capabilities require evolved skill sets. Today's data scientists might need quantum computing knowledge tomorrow. Current ML engineers may require expertise in brain-computer interfaces. Invest in continuous education, supporting employees pursuing advanced degrees, attending conferences, and experimenting with emerging tools. Adjust recruiting profiles; hiring for learning agility rather than just current technical skills.

Regulatory Anticipation: AI regulation lags technology but arrives inevitably. Companies preparing for future AI must anticipate governance evolution. The EU's AI Act provides a template other jurisdictions will likely follow. Organizations building ethical AI frameworks, audit trails, and explainability capabilities today avoid expensive retrofitting when regulations mandate these features tomorrow.

Scenario Planning Discipline: Regular scenario planning exercises explore multiple AI evolution pathways: gradual capability improvement versus sudden breakthroughs, open versus proprietary development models, centralized versus distributed architectures. Each scenario suggests different strategic implications. One bank developed contingency plans for three scenarios: AI commoditization (where competitive advantage shifts from algorithms to data), AI concentration (where a few players dominate), and AI democratization (where capabilities become universally accessible).

Platform Evolution Strategy: Today's AI platforms must evolve continuously rather than face periodic replacement. Cloud providers like AWS, Azure, and Google Cloud continuously integrate new AI capabilities without disrupting existing services. Enterprise AI platforms need similar evolution mechanisms: backward compatibility guarantees, graceful capability introduction, and clear migration pathways for legacy systems.

Ecosystem Positioning: Future AI value may flow through ecosystems rather than individual companies. Strategic positioning within emerging AI networks, whether industry consortiums, open-source communities, or commercial partnerships, determines access to next-generation capabilities. One automotive company participates in multiple autonomous vehicle consortiums, ensuring visibility into competing technical approaches while building relationships for future collaboration.

Ethical Leadership: As AI capabilities expand, ethical considerations multiply: ranging from bias and privacy to transparency and accountability. Organizations that lead on AI ethics today help shape tomorrow's standards and public expectations. Ethical leadership goes beyond regulatory compliance; it builds trust, which becomes a strategic asset when deploying powerful new technologies. Studies show that companies embedding ethical governance frameworks covering areas like fairness, explainability, and data protection are better positioned to scale AI responsibly and gain stakeholder confidence. This trust enables faster adoption, smoother integration, and greater resilience when navigating complex societal and regulatory landscapes.

Infrastructure Modernization: Legacy technical debt constrains AI evolution more than any other factor. Continuous infrastructure modernization with cloud migration, API enablement, and microservices architecture creates the foundation for integrating future AI capabilities. One manufacturer's three-year infrastructure overhaul seemed excessive until it enabled rapid deployment of computer vision, natural language processing, and predictive analytics across their operations.

Innovation Culture Cultivation: Technology evolves faster than organizations. Cultural readiness for continuous change matters more than specific technical preparations. Companies fostering experimentation, celebrating intelligent failure, and rewarding adaptation build organizational muscles for whatever emerges. This cultural foundation explains why some organizations thrive through multiple technology disruptions while others excel in one era but stumble in the next.

The journey from AI strategy to sustained value creation requires a holistic organizational transformation. This transformation must span every dimension we've explored: aligning strategy with execution, embedding continuous optimization, and preparing systems and culture for future evolution. Only by integrating these elements can organizations move beyond isolated successes to build scalable, resilient AI capabilities that deliver lasting impact.

SUMMARY

The journey from AI strategy to execution reveals a truth that separates the four percent of companies achieving significant returns from the 96 percent trapped in pilot purgatory: implementation is transformation, not installation. Throughout this chapter, we've dismantled the dangerous myth that AI success follows from technological sophistication alone. Instead, sustainable value emerges from orchestrating organizational change across multiple dimensions simultaneously.

The framework we've explored forms an integrated system, not a sequential checklist. Each element reinforces the others. Strategic alignment without metrics leads to drift. Metrics without resources starve promising initiatives. Resources without scaling capabilities waste investment on isolated successes. Scaling without continuous learning produces obsolescence. And all these capabilities mean little if organizations can't adapt to AI's next evolutionary leap.

Several meta-lessons emerge from successful AI transformations:

Execution Velocity Beats Perfect Planning: The most detailed roadmaps rarely survive first contact with organizational reality. Winners maintain strategic direction while adapting tactical execution rapidly. They prize learning speed over planning precision, using metrics and feedback loops to course-correct continuously rather than pursuing rigid blueprints.

Portfolio Thinking Trumps Project Management: Traditional ROI calculations applied to individual AI projects miss the transformational forest for the efficiency trees. Successful organizations manage AI as a portfolio, accepting intelligent failures as the price of breakthrough innovation while ensuring overall value creation. This venture capital mentality enables bolder moves than project-by-project justification permits.

Cultural Change Determines Technical Success: The most sophisticated algorithms fail in cultures that resist change, distrust automation, or lack data literacy. Conversely, organizations with modest technical capabilities but strong change management often outperform technically superior but culturally rigid competitors. The human dimension deserves equal investment with the technical foundation.

Platform Beats Point Solutions: Organizations scaling multiple AI successes invariably build platforms: shared infrastructure, reusable components, common governance frameworks. This platform foundation accelerates subsequent initiatives while ensuring consistency. The investment seems excessive for single projects but proves economical across the portfolio.

Leadership Alignment Multiplies Impact: When C-suite executives view AI as "the technology team's project," initiatives remain marginal regardless of technical merit. Transformational impact requires leadership alignment from board to front line. This doesn't mean executives become data scientists, but they must understand AI's strategic implications deeply enough to champion change.

Continuous Evolution Sustains Advantage: AI transformation has no finish line. The capabilities, organizations, and cultures we build must evolve continuously. Static excellence becomes competitive disadvantage as the field advances. Build organizational metabolism: the ability to absorb new capabilities, shed outdated approaches, and adapt structures dynamically.

The road ahead promises both unprecedented opportunity and existential challenge. AI capabilities will advance more in the next five years than the previous fifty. Organizations that master the execution framework outlined here position themselves not just to survive but to shape this transformation. They'll build the adaptive capacity to harness whatever emerges from AI's accelerating evolution.

Yet amid this technical revolution, a deeper truth persists: AI remains a tool wielded by human judgment for human purposes. The organizations that remember this while executing with excellence will write the next chapter of business history. They'll achieve what every transformational technology promises but few deliver: fundamental advancement in how humans create value, solve problems, and improve lives.

10

TAMING THE ALGORITHM: BUILDING HUMAN-FIRST AI ORGANIZATIONS

Organizations that lose sight of their human core while chasing AI capabilities create hollow victories. They optimize processes only to find their people disengaged, deploy sophisticated systems that employees circumvent, and achieve efficiency gains at the cost of institutional wisdom. The challenge facing Global 2000 companies isn't mastering AI technology; it's ensuring that technology amplifies human judgment rather than replacing it.

Throughout the preceding chapters, we've witnessed this pattern repeatedly. Technology alone guarantees nothing. A revealing study found that 63 percent of organizations cite human factors as their primary AI implementation challenge. The disconnect runs even deeper. While 96 percent of C-suite leaders celebrate AI's productivity promise, 77 percent of employees report that AI has actually increased their workloads. This phenomenon of "fake productivity" emerges when organizations push technology without preparing their people, creating more friction than flow.

Consider the retail case study from early in this book. When a rushed AI deployment created additional work for store employees tasked with correcting system errors, the technology meant to liberate became a burden. The organization achieved computational sophistication while losing operational effectiveness. In contrast, when companies approach change with people at the center, AI becomes a catalyst rather than a constraint. Business leaders and researchers increasingly recognize this imperative. As Mercer's 2024 talent trends study observed, "raising productivity through AI...requires intentional, human-centric work design." Companies must reimagine processes, roles, and workflows so that technology genuinely supports rather than supplants human capability.

The path forward synthesizes the lessons from our AI transformation framework into a blueprint for enduring, people-focused change. We'll examine how to align AI initiatives with organizational values and mission, creating technology that serves purposes everyone understands. We'll explore methods for cultivating cultures that encourage learning, collaboration, and psychological safety during technological transitions. And we'll investigate how leadership styles and governance mechanisms must evolve to champion human-centric principles from boardroom to break room.

Most critically, we need to reimagine employee empowerment through strategic upskilling, role redefinition, and systemic inclusion. Employees shouldn't compete with AI; they should thrive alongside it. This requires human-centered design of AI systems and processes, ensuring that tools emerge from genuine user needs rather than technical possibilities. Since transformation isn't a singular event but an ongoing evolution, we'll also consider how to sustain and scale these practices as AI capabilities advance.

By chapter's end, you will possess a comprehensive understanding of how to cultivate AI-powered organizations that never lose sight of their people. The sections ahead illuminate each dimension of this challenge, from values and culture to leadership and design, demonstrating how AI transformation can pivot around human strengths rather than despite them. We start by showing why keeping humans at the center of AI isn't just a nice idea; it's a smart business move.

TRANSFORM TECHNOLOGY, AMPLIFY HUMANITY

A human-centric AI organization fundamentally treats AI as capability augmentation rather than human replacement.

This mindset shift cascades through every organizational level, from strategic planning to process design. Instead of asking, "Which jobs can we automate with AI?" ask, "Which tasks can we streamline so our people focus on higher-value work?" Augmented intelligence becomes the operating philosophy. Employees retain ultimate decision-making authority while leveraging AI insights to enrich judgment calls. This approach preserves work's purpose and dignity, motivating staff to partner with AI tools rather than resist them.

Human-centered AI adoption actively creates competitive advantage beyond pitfall avoidance. Organizations successfully blending human and AI strengths consistently achieve:

Higher adoption and ROI: Engaged employees use AI systems more effectively, ensuring technology delivers promised benefits rather than gathering digital dust.

Greater innovation: When AI automation liberates people from routine drudgery, they redirect energy toward creative, strategic endeavors that spark new ideas and improvements.

Stronger trust and ethics: Maintaining humans "in the loop" provides oversight preventing AI errors or biases, reinforcing stakeholder confidence.

Employee retention and growth: Rather than alienating the workforce, human-centric strategies signal investment in people's futures. Employees stay and grow when they see technology enhancing rather than eliminating their roles.

Companies must embrace "both/and" thinking (leveraging AI's computational power alongside human judgment) instead of forcing "either/or" trade-offs. This philosophy threads throughout our AI transformation framework, acknowledging that while AI processes data at superhuman scale, humans provide context, intuition, and ethical discernment that machines lack. Combining these complementary strengths produces superior decisions and outcomes than either could achieve independently. Moreover, centering discussions on humans addresses legitimate workforce anxieties, demonstrating leadership's commitment to employee wellbeing throughout the AI transition. This foundation of trust becomes essential cultural bedrock.

Recentering on humans thus establishes the cornerstone for every subsequent AI integration aspect. It reminds us that technology's value ultimately measures through its impact on people: employees, customers, and society. The next step ensures all AI efforts anchor firmly to the organization's core purpose and values. The following section examines how aligning AI with company values creates an ethical and effective implementation beacon.

ANCHOR AI TO ORGANIZATIONAL PURPOSE

Technology carries no inherent values, but its application can either reinforce or undermine an organization's core principles. Before examining technical capabilities, ask the fundamental questions: How does this AI project serve our purpose? Does it uphold our values?

When AI initiatives visibly advance an organization's mission like improving customer experience, enhancing safety, or elevating quality, stakeholder buy-in accelerates. When they conflict with stated values, resistance is inevitable.

According to public reporting, ethics frameworks must progress from aspirational statements to practical implementation. Leading organizations are publicly documented as codifying principles through charters and governance boards. For instance, reports highlight how Mayo Clinic operationalized its commitment to "augment, not replace, human care" by requiring every AI tool to show improved patient outcomes and enhanced clinician effectiveness. Similarly, public sources detail IBM's stance on societal responsibility, noting its decision to discontinue general-purpose facial recognition to better align with human rights principles.

Financial services illustrate the cost of misalignment. Upstart faced scrutiny over lending bias and responded with bias audits and transparency measures under CFPB oversight. According to public reporting, Microsoft has institutionalized oversight through its Responsible AI Standard and Sensitive Use reviews, with committees empowered to halt deployments that breach ethical thresholds.

Values alignment now spans diversity, equity, inclusion, and sustainability. According to public reports, Amazon chose to discontinue its AI hiring tool following concerns about bias, illustrating the consequences of overlooking these priorities. In contrast, public reporting highlights that Apple's approach to privacy-focused AI design has reinforced brand trust and deepened customer loyalty.

Employees and customers reward organizations that embed ethics into technology. According to public reports, BMW's approach of ensuring that "technology serves people" has guided its AI adoption in manufacturing. This strategy, which prioritized not displacing workers, was instrumental in converting initial skeptics into advocates. Research confirms the payoff: nearly 70 percent of employees say values alignment is a deal-breaker when choosing an employer.

Grounding AI in purpose and values provides clarity for tough trade-offs and sets boundaries for acceptable applications. It is not a one-time declaration but an ongoing dialogue, essential for sustaining trust as technology and expectations evolve. Even with clear values and ethics frameworks established, AI transformation falters when organizational culture resists change. Culture represents the shared mindsets and behaviors defining "how we do things here."

BUILD CULTURE WHERE PEOPLE AND AI THRIVE TOGETHER

Successful AI adoption requires cultures encouraging people-first adaptation: employees must feel safe, motivated, and empowered to integrate new technologies into daily workflows.

Trust and transparency are the foundation. Employees embrace AI when they understand why it's being introduced and how it will affect their roles. Leaders who communicate openly, acknowledge concerns, and pair promises with safeguards reduce fear and build confidence.

Real-world examples show what this looks like in practice.

According to public reporting, Starbucks positions its "Deep Brew" platform as a tool designed to enhance barista roles by optimizing staffing, inventory, and personalization. This approach allows store teams to dedicate more time to customers, rather than focusing on repetitive tasks. Reports highlight that the company's language and leadership behaviors stress augmentation over replacement, which has been instrumental in integrating AI as an essential part of service delivery. In a similar vein, public sources indicate that IKEA introduced its "Billie" AI assistant in customer support while making a public commitment to upskill thousands of employees into higher-value advisory positions. By leveraging AI as both an efficiency driver and a catalyst for career advancement, IKEA, as documented in these reports, transformed initial apprehension into positive momentum.

Cultures that thrive with AI share common traits. They keep communication open and continuous, explaining the "why" behind changes and inviting questions. They involve employees in shaping solutions through pilots and design workshops, because people support what they help create. They encourage safe experimentation, treating smart failures as learning opportunities rather than career risks. And they lead with empathy, listening to anxieties about job security and responding with retraining pathways and visible commitments.

When these elements come together, AI feels less like a top-down mandate and more like a natural extension of how the organization improves. Employees feel respected and empowered, curiosity replaces dread, and adoption accelerates.

Shaping such cultures requires consistent reinforcement. Symbols and behaviors matter profoundly. When executives personally use new AI tools and share learning experiences, powerful messages resonate. Similarly, adjusting performance metrics to value learning and collaboration beyond short-term output nudges culture appropriately. Over time, people-first cultures become self-reinforcing: AI successes breed confidence, spawning further experimentation and achievement. However, culture change occurs within leadership context. The next section examines leadership and governance's pivotal role in modeling and sustaining human-centric approaches.

LEAD WITH HUMANITY IN THE AGE OF ALGORITHMS

Executive sponsorship are some of the top factors in successful, human-centered AI transformation.

Survey evidence reinforces a critical pattern: 43 percent of organizations attribute AI adoption failure to insufficient executive sponsorship. Leaders set the organizational tone. When they champion human-centric visions for AI, fund people-focused initiatives, and hold themselves accountable for transformation outcomes, the enterprise follows. When leaders take a hands-off posture or treat AI as a narrow IT project, momentum drains from even the best-intentioned efforts.

Human-centric leadership starts by articulating a clear, credible, and inspiring vision for how AI elevates the organization, one that moves beyond profit and efficiency to connect with purpose and with tangible benefits for employees and customers. According to public reporting, DBS Bank CEO Piyush Gupta has consistently positioned AI as a means to free employees from low-value tasks, enabling them to focus more on advisory conversations, service recovery, and creative problem-solving for clients. These reports indicate that he recognized the challenges of adapting to new tools and workflows and pledged to provide training for employees impacted by these changes, demonstrating both empathy and shared accountability. This publicly documented approach helps frame AI as an opportunity rather than a threat, offering managers a narrative they can reinforce in day-to-day decisions.

The message only sticks when leaders lead by example. AI adoption cannot be delegated; executives have to change how they work. DBS senior leaders enrolled in data and AI learning alongside middle managers and frontline supervisors, underscoring that capability building is everyone's job, including theirs. Leaders began using internal analytics and AI-enabled dashboards to inform decisions and shared what they learned, false starts included. When employees see their leaders experimenting, admitting what they don't know, and still anchoring choices in human values, trust grows - and so does the willingness to try new methods.

Leadership in the AI era also requires a different mindset. AI introduces uncertainty and ethical ambiguity. Effective leaders practice humility in the face of imperfect models, curiosity about how systems reach recommendations, and inclusivity in how decisions are made. They probe AI outputs rather than accepting them at face value. They invite debate and welcome dissent when interpreting algorithmic insights to avoid blind spots. They resist the "AI knows best" trap. This reflective posture aligns with the principle of being data-informed rather than merely data-driven: values and human judgment remain the final arbiters. And rather than concentrating judgment in a few technical experts, wise leaders cultivate broad trust networks so people at every level feel safe raising concerns and suggesting improvements.

Governance is the companion to tone and mindset: the scaffolding that keeps AI transformation on a human-centric trajectory. Leading companies stand up cross-functional bodies to guide AI strategy, assess risks, and monitor impact. Microsoft's Aether committee and Office of Responsible AI work with product, research, legal, HR, and customer-facing teams to operationalize principles through policy and process. Requirements such as human oversight for material decisions, impact assessments for sensitive use cases, and systematic testing for bias convert aspiration into practice. These structures don't slow innovation; they channel it toward outcomes that sustain trust.

Board oversight is the final safeguard. Treating AI as a strategic priority on par with cybersecurity or financial controls reframes executive incentives beyond short-term ROI. Directors are asking sharper questions about human capital and stakeholder effects: How are we reskilling people whose roles are changing? What ethical guardrails govern development and deployment? How will we measure AI's impact on engagement, inclusion, and customer trust? Some boards have added AI expertise or convened external ethics advisors to strengthen independence of judgment, recognizing that balancing speed with responsibility is now a fiduciary concern.

Human-centric leadership, then, is both stance and system. Leaders champion and model the change; governance embeds checks and balances to ensure technology never outruns the organization's capacity to manage it responsibly. Under this kind of stewardship, employees gain confidence that transformation will proceed thoughtfully. With leadership and governance in place, the center of gravity can shift to the organization's engine - its people - and the skills, roles, and ways of working that unlock an AI-enriched workplace.

REDEFINE WORK: EMPOWERING PEOPLE IN THE AI ERA

When AI accelerates faster than human skills, the companies that win are those that turn capability building from an initiative into an operating system.

Fifty-eight percent of executives say technology is advancing faster than their firms can retrain workers, and fewer than half believe their current talent models can meet near-term demand. Bridging this gap is existential. Without capable, confident workforces, even the most sophisticated AI systems gather dust. Human-centric AI organizations therefore invest deeply in reskilling, upskilling, and role redefinition, ensuring employees don't just survive transitions but thrive within them.

Journeys that work typically begin with a frank skills baseline and bold learning investments. According to publicly available reports, when Amazon scaled its AI-enabled robotics, the company introduced a Department-of-Labor–registered Mechatronics & Robotics Apprenticeship program. This initiative combines 12 weeks of paid classroom instruction with approximately 2,000 hours of mentored, on-the-job training, and guarantees participants full-time technical maintenance roles upon completion. Public sources indicate that this apprenticeship is part of Amazon's broader upskilling commitment, which has involved investing more than $700 million and engaging hundreds of thousands of employees worldwide, including front-line associates preparing for higher-skilled positions. By treating learning time as real work and linking it to recognized credentials and opportunities for internal mobility, Amazon transforms development from an extracurricular activity into a clear career pathway.

Role evolution sits at the heart of the strategy. Rather than eliminating jobs AI can now perform, redesign roles to emphasize uniquely human capabilities. According to public reporting, HSBC utilizes generative AI to assist with credit-analysis write-ups and expedite certain aspects of the credit application process. These sources indicate that analysts are able to devote more time to nuanced risk assessment and personalized client conversations, while AI systems handle the assembly and summarization of the data-intensive inputs. This direction aligns with broader industry guidance: credit-risk teams are adopting gen-AI tools for synthesis and drafting, while keeping human oversight for interpretation and final decisions. The result are analysts whose craft shifts from compilation to counsel.

Sustaining these shifts requires a culture of continuous learning that goes beyond courses. According to public reporting, ING established a Global Analytics Academy to enhance data literacy throughout the bank, which involved assessing, training, and certifying more than 5,000 specialists, while also introducing tens of thousands of colleagues to analytics concepts and "business translator" skills. Critically, the learning happens in the work: cross-functional squads pair domain veterans with technical experts so knowledge flows both ways and solutions improve through frontline feedback.

Empowerment is psychological as much as it is technical. Employees need confidence and permission to use AI. According to public reporting, Bank of America has taken a two-pronged approach: "The Academy" delivers high-tech, high-touch development through immersive simulations utilizing AI and VR, while internal assistants like Erica for Employees—now used by more than 90 percent of staff—help reduce friction in daily tasks. When people see peers succeeding with AI and feel backed by training and tools, they are far more likely to experiment and adopt.

Done thoughtfully, implementation can usher in a genuine upgrade in the quality of work. Workers report spending less time on rote tasks and more on judgment-heavy, creative, and client-facing activities, especially when companies match upskilling with clear mobility guarantees. According to public reporting, Amazon's apprenticeship model culminates in placement into higher-skill maintenance roles, providing assurance that lowers fear and unlocks engagement. Confidence grows when organizations signal, through policy and investment, that willing learners won't be left behind.

Reimagining work also surfaces entirely new roles. "Data translators" who bridge business and technical teams are now common in skills academies like ING's; many companies are formalizing AI ethics leadership to guide responsible use; and "automation coordinators" increasingly stitch together human and machine workflows. To widen access to opportunity, firms are building internal talent marketplaces. According to public reporting, Unilever's AI-powered FLEX Experiences has helped redeploy thousands of employees during periods of demand shocks and increased productivity by uncovering hidden skills among more than 100,000 staff members. Similarly, public sources indicate that Schneider Electric's Open Talent Market has seen widespread adoption, resulting in tens of thousands of project and mentor matches and facilitating large-scale internal mobility.

The underlying theme: human-centric organizations treat people as their greatest AI-era asset, not a cost to be minimized. Investment in skills and growth closes technical gaps while building trust and loyalty. When employees believe the company is investing in their future, they embrace new tools with energy. The payoff is an agile, continually evolving workforce that unlocks AI's potential. With talent empowered, the next imperative is ensuring the systems themselves are designed for usability, ethics, and day-to-day support of human work: the focus of the following section.

DESIGN AI FOR HUMANS

Too often, companies deploy AI tools that technically function but fail because they weren't designed with end-users in mind. Clunky interfaces, opaque outputs, and weak workflow integration can sink even the smartest systems. When tools are hard to use or live outside daily rhythms, adoption stalls regardless of underlying sophistication. Avoiding these traps requires a human-centered approach that begins with deep user research, continues through iterative prototyping, and prioritizes usability, transparency, and trust. Solutions should feel intuitive. For example, consider how Target expanded machine learning for product availability and then put guidance in the hands of store teams through familiar devices, introducing Store Companion as a generative AI assistant that streamlines procedures and answers process questions in context. The objective was simple: fewer app switches, clearer next actions, and support inside the handheld workflows associates already use. Leaders described the impact as removing friction and giving teams more time with guests, which is how adoption takes root.

The people who will live with a system every day should shape it from the earliest sketches. Organizations that run design workshops, observe real work, and pilot with frontline teams surface the moments that matter. In retail operations, for example, embedding recommendations directly into handheld dashboards at the moment of task execution beats sending users to a separate analytics portal. The same logic applies in service, underwriting, and plant operations. Frequent pilots and short feedback loops let teams reduce cognitive load, simplify language, and tune timing so that suggestions arrive when they are wanted, so they don't get in the way.

Clinicians, analysts, and supervisors trust systems more when they can see why a recommendation emerged. In medical imaging, studies of diabetic retinopathy detection show how heatmaps and class-activation visualizations help clinicians understand which features drove a model's conclusion, which reduces the black-box effect and supports joint decision making. Reviews of explainable AI in healthcare reinforce the same point, highlighting how techniques like Grad-CAM, SHAP, and LIME can make advanced models more interpretable in practice.

Teams should define who acts, when, and with what level of autonomy. In finance, human-in-the-loop toolkits such as ALARM (Analyst-in-the-Loop Anomaly Reasoning and Management) support anomaly detection with interactive triage, analyst explanations, and rule updates so that alerts become actions only after a person reviews and authorizes them. The result is a system that learns from expert feedback while preserving accountability for high-stakes decisions.

AI succeeds when it connects to the tools and data people already rely on and when outputs flow into downstream steps without extra keystrokes. In customer work, Salesforce has embedded Einstein capabilities inside CRM screens so that next-best actions, summaries, and automations sit where sales and service teams already spend time, and admins can generate and refine flows rather than bolt on another application. In frontline retail, Walmart's associate app now hosts AI tasking and translation features, reducing shift-planning effort and supporting natural language interactions on the devices associates carry. These are examples of meeting users where they are instead of asking them to shuttle among systems. Monitor real usage, gather feedback, and update systems as needs evolve. Walmart's rollout explicitly builds iterative feedback loops into associate tools so performance improves over time, which is how trust compounds and interfaces stay aligned with reality on the ground.

When organizations design AI for and with the people who will use it, technology becomes a natural extension of the job. Tools feel intuitive, useful, and trustworthy because they respect human judgment and fit actual workflows. That is the practical definition of empathy in design. It signals investment not only in culture and skills but also in the day-to-day experience of work. With these foundations in place, we can focus on sustaining and scaling human-centric ecosystems as AI capabilities and business needs continue to evolve.

SUSTAIN THE HUMAN-AI PARTNERSHIP FOR LONG-TERM SUCCESS

The real test of human-centric AI begins after launch, when new habits either take root or quietly slip back to old routines. After the first waves of implementation, the work shifts to sustaining and steadily improving the new ways of working. This phase determines whether early gains compound or erode. Leaders and change agents have to ensure that the practices and mindsets established during rollout become part of the company's DNA. The objective is to make adaptability and human-centric thinking enduring capabilities that help the organization navigate today's changes and tomorrow's as well.

Organizations improve what they measure, so human-centric objectives need explicit tracking. That means pairing quantitative indicators such as engagement, adoption, and outcome metrics with qualitative signals from focus groups and open suggestion channels. According to publicly available reports and sources, retailers such as Walmart have demonstrated how to close the loop by integrating AI directly into the associate app, actively monitoring usage and task outcomes, and iterating features based on frontline feedback rather than requiring employees to adjust their work to fit the tool. Early results reported publicly—including time saved in shift planning and ongoing translation improvements—highlight an approach focused on listening and refining within the flow of work. Similarly, public reporting on University of Michigan Health-West's introduction of ambient documentation AI indicates that leaders linked adoption to regular pulse checks on clinician workload and experience, then adapted training and support based on collected feedback. These positive outcomes were sustained because leaders treated ongoing feedback as an essential part of their operating rhythm.

Cross-functional AI councils and steering committees need to keep meeting, scanning new use cases, and resolving trade-offs. According to publicly available reports and sources, HSBC has embedded AI

Review Councils across the group and aligned lifecycle management with risk frameworks, ensuring that oversight remains active as new generative and predictive applications are introduced. Public reporting also indicates that governance includes ethics reviews and model audits that recur over time. JPMorgan's approach to bias evaluation for risk models, as described in public sources, highlights how large institutions are developing mechanisms to detect and mitigate disparate impact as systems evolve, rather than treating fairness as a one-time check.

According to public reporting and publicly available sources, instead of relying on one-time training that coincides with system launch, companies are increasingly making AI literacy and tool proficiency a core part of onboarding and career development. For example, DBS has put this approach into action with its Future Tech Academy and bank-wide upskilling initiatives in areas such as AI and data. Similarly, ING has implemented a Global Analytics Academy that has enhanced data literacy at scale and certified thousands of employees via role-specific learning paths. These programs are designed to evolve, ensuring that capability building stays ahead of the technology curve.

Publicly reported examples also show that organizations reinforce desired behaviors by recognizing ideas for improvement, updating performance conversations to include effective human-AI collaboration, and broadcasting stories that illustrate successful practices. As reported in public sources, Unilever's FLEX talent marketplace has generated a steady flow of such stories, including redeployments during demand shocks and employees discovering new roles through internal gigs. Sharing these successes has helped embed the understanding that smart technology and human initiative go hand in hand. Schneider Electric's Open Talent Market, as described in public reports, has produced similar peer-driven advocacy, with mentors and hiring managers modeling behaviors the company aims to scale.

Ongoing changes in capabilities, regulations, and societal expectations are frequently noted in public reporting. Organizations that maintain resilience adapt policies and practices proactively, rather than reacting to external pressures. HSBC's continuing efforts to operationalize responsible AI—including explainability and oversight of sensitive use cases—have been highlighted in publicly available reports as examples of how living governance systems can absorb new technologies like generative AI while maintaining stability. These sources indicate that human-centric habits accelerate adaptation, as teams engaged in open dialogue more readily identify skill gaps and ethical risks and address them promptly.

Public sources recommend treating human-centric AI management as an ongoing journey requiring regular reflection, even when adoption appears strong. Keeping "people impact of AI" as a standing agenda item, testing whether employees continue to embody guiding principles, and exploring where new innovation can further the mission have all been emphasized in recent public research. For instance, Mercer's 2024 Global Talent Trends reports that 58 percent of executives believe technology is moving faster than their firms can retrain staff, and only 47 percent feel current talent models can meet short-term needs. McKinsey's 2025 research indicates that while nearly all companies are investing in AI, only one percent consider themselves mature, with leadership often cited as a key barrier to scaling outcomes. Gartner's 2024 polling of HR leaders, as publicly reported, notes persistent challenges in leveraging skills effectively and building future capabilities confidently, underscoring that technology alone does not transform work.

According to publicly available reports, the strongest evidence that change has taken hold is when employees themselves advocate for human-centric practices. Internal marketplaces and communities of practice foster this advocacy. Both Unilever's FLEX and Schneider's Open Talent Market, as documented in public sources, illustrate how, even a year or more after rollout, peers mentor one another, surface misalignments, and drive improvements from the ground up. This kind of self-correction diminishes the need for top-down enforcement and helps keep systems aligned with their intended purpose.

Documenting results such as faster cycle times, higher retention, or improved customer outcomes, and tying those gains to human-centric methods, builds conviction to continue the path. Bank of America has publicly reported scale metrics that link AI adoption to employee productivity improvements, including more than 2.5 billion client interactions with Erica and internal usage of AI assistants by over 90 percent of employees. Evidence at that scale helps leaders and teams see why disciplined, people-first implementation matters.

Through ongoing measurement, reinforcement, adaptation, and celebration, human-centric AI becomes a dynamic capability rather than a static end state. Executives continue to worry about keeping people inspired and skilled for new tools, while HR leaders acknowledge gaps in how technology actually changes work. Companies that get this right treat human-centric change management as a permanent strategic system. They accept that technology will keep advancing, and they place their competitive bet on how well they can keep aligning those advances with human ingenuity and need.

SUMMARY

In this chapter we distilled the AI transformation into a practical blueprint for building organizations where technology amplifies human potential. We began with first principles: people belong at the center of AI, not for sentimental reasons but because trust, adoption, and judgment decide outcomes. Anchoring initiatives in purpose and values gives leaders an ethical and strategic compass, so technology serves meaningful ends rather than becoming the end itself.

We then examined the ground conditions where AI either flourishes or falters. Environments that reward learning, transparency, and inclusion create the confidence required to experiment, to question model outputs, and to improve the work. Leadership is the catalyst.

Executives and boards must model human-centric behaviors, set simple rules for human oversight, and install governance that turns principles into process.

Companies that invest in reskilling and redesign roles to elevate judgment, creativity, and relationship work unlock an "augmented" workforce. Tools only become tools when they fit the job, so systems should be designed with and for users, with emphasis on usability, explanation, and seamless integration into everyday workflows.

Finally, we made sustaining the norm rather than the exception. Continuous improvement, recurring ethics reviews, and ongoing capability building convert one-time wins into durable competencies. When measurement, feedback, and learning are routine, organizations retain their human core while keeping pace with technological change.

Chapter takeaways:

- Trust, judgment, and everyday adoption by the workforce unlock value beyond model accuracy.
- Values anchor direction: grounding AI in purpose provides the compass for tough trade-offs and keeps technology serving meaningful ends.
- Environments that reward learning, transparency, and inclusion accelerate adoption and improve decision quality.
- Executives and boards must model the behaviors, fund people-first investments, and sustain active governance.
- Reskilling and role redesign elevate judgment, creativity, and customer connection across roles.
- Intuitive, explainable, seamlessly integrated tools become daily aids rather than add-ons to the job.
- Continuous measurement, recurring ethics reviews, and ongoing learning convert one-time wins into durable capabilities.

With the organizational blueprint in place, the lens now widens. The same human-centric principles that reshape a single enterprise also guide how leaders engage the wider system that AI is transforming. Chapter 11 examines how to take this playbook beyond the firm: shaping industry standards, navigating emerging regulation, strengthening ecosystem partnerships, and building societal trust so that AI's benefits compound for customers, employees, and communities.

11

NAVIGATING THE BROADER AI LANDSCAPE

As you progress in their AI transformation journey, the focus expands beyond internal change management to include the broader external environment. While previous chapters focused on internal strategies within the enterprise's walls, we now turn outward to examine how global forces, industry trends, and external stakeholders shape the way Global 2000 companies lead and adapt in an age of AI dominance.

The Global AI ecosystem comprises interconnected networks of technology providers, research institutions, talent pools, and markets that collectively drive innovation. Unlike earlier technological eras, AI development unfolds across North America, Europe, and Asia simultaneously, creating both opportunities and challenges for multinational enterprises.

In this chapter, you will discover how major regions and countries contribute to AI innovation, which industries lead the charge, and what this interconnected landscape means for organizations navigating across borders. Beyond technology adoption, leaders must prepare for the regulatory and ethical challenges on the horizon, as governments worldwide race to create frameworks ensuring responsible AI use. The discussion illuminates how different regions approach AI governance and what global companies must do to comply with and shape these emerging rules.

Emerging AI technologies and trends demand strategic attention from C-suite leaders. Beyond familiar AI applications, innovations like generative AI, autonomous systems, and multi-modal learning stand ready to disrupt industries. Organizations that understand these technological trends can better anticipate opportunities and threats, ensuring they maintain competitive advantage rather than falling victim to the next wave of AI advancements.

The chapter addresses preparation for external AI-driven disruption, recognizing that innovation alone provides insufficient protection. Companies must cultivate organizational agility to respond when competitors or new entrants leverage AI in game-changing ways. Consider strategic partnerships and alliances as tools for accelerating learning and innovation, as no single company can master the full spectrum of AI capabilities independently.

Leading enterprises shape their competitive environment by influencing industry standards and best practices. Global 2000 companies possess both the opportunity and responsibility to guide AI deployment across their industries through participation in international standards, policy dialogues, and industry consortia focused on AI ethics and safety.

Finally, this chapter culminates in a comprehensive view of leadership in an AI-dominated world. We synthesize global awareness, regulatory navigation, technological foresight, external readiness, collaboration, and industry leadership to describe what effective AI-era leadership requires.

NAVIGATE THE GLOBAL AI POWER STRUCTURE

Global companies now operate in an AI environment that looks very different from just three years ago. What was once a race led by a handful of tech giants has become a multipolar system where innovation, regulation, and investment flow across continents at unprecedented speed.

The United States remains the epicenter of AI scale and infrastructure. American firms such as Microsoft, Google, Amazon, and Meta continue to dominate cloud platforms and frontier model development, while NVIDIA's rise to a market cap above $4 trillion in 2025 underscores how control of computing power has become the ultimate advantage. U.S. policy has shifted from safety-first principles to a deregulatory, pro-innovation stance: new directives focus on removing regulatory barriers and accelerating American AI leadership rather than imposing broad safety and transparency mandates. For global enterprises, this means U.S. partnerships often provide early access to cutting-edge tools without the heavy compliance burden that characterized prior approaches. The U.S. AI market alone is projected to exceed $80 billion in 2025, and venture funding for AI startups has surged past $50 billion annually, fueling a wave of acquisitions and partnerships.

Europe has taken a different path, positioning itself as the regulatory anchor and a leader in industrial AI. The EU AI Act is now in force, introducing phased obligations for risk management, transparency, and post-market monitoring. Companies operating in Europe must design systems that meet these requirements from day one. At the same time, Europe continues to excel in manufacturing, automotive, and healthcare AI, making it a hub for specialized applications and ethical frameworks. The UK maintains a lighter, principles-based approach for now, but legislative proposals signal that stricter oversight may be coming.

Based on information reported in public sources, Asia is recognized for its scale, speed, and ambition in AI. Public reporting indicates that China remains a significant force despite export controls on advanced chips, increasing investment in domestic silicon and AI platforms. According to publicly available data, Baidu, Alibaba, and Tencent have launched large-scale generative AI models for search, commerce, and enterprise services, while Huawei advances its Ascend chips to reduce reliance on U.S. technology. India's national AI mission, as widely reported, is developing a compute platform targeting over 10,000 GPUs to support startups and research, with Infosys and TCS integrating generative AI into global IT services. News sources highlight Japan's SoftBank investments in robotics and AI ventures, while NEC and Fujitsu concentrate on trusted AI for government and enterprise. Public information shows South Korea's Samsung and SK Hynix lead in high-bandwidth memory for AI chips, and Naver recently introduced HyperCLOVA X for Korean-language generative AI. Singapore is noted in public reports for fostering experimentation through regulatory sandboxes and partnerships with global cloud providers. While these markets present considerable talent and growth opportunities, publicly available sources suggest organizations must carefully navigate local compliance and supply chain dynamics.

Recent news coverage identifies the Middle East as an emerging center of AI activity. For example, Abu Dhabi's G42 partnership with Microsoft and Saudi Arabia's proposed $40 billion AI fund have been publicly announced as part of sovereign efforts to build regional AI ecosystems. These initiatives, reported in the media, are creating new hubs for data centers, chips, and venture investment, offering companies alternative infrastructure and partnership opportunities outside traditional markets.

Publicly available information shows open-source AI is accelerating global diffusion. Meta's Llama 3.1 and Mistral's recent releases, as described in news articles and press releases, enable enterprises to develop custom solutions without depending solely on proprietary systems. This is particularly relevant in regions with stringent data laws, such as Europe, where organizations require more control over AI operations within their environments. According to public sources, the combination of advances in generative and multimodal systems and open models is reshaping strategies for compliance, cost, and technological sovereignty.

The defining feature of this landscape is interdependence. A breakthrough in California can be fine-tuned in Paris and deployed in Singapore within weeks. Global companies cannot afford to think locally. They need strategies that combine U.S. innovation, European compliance, Asian scale, and Middle Eastern capital. That means mapping partners, securing compute supply, blending open and closed models, and building governance that works across jurisdictions.

Worldwide AI spending is projected to exceed $200 billion in 2025, with double-digit growth continuing through 2026. Companies that tap into this global network through partnerships, acquisitions, and shared standards position themselves to lead.

With this global picture in mind, the next challenge is regulation and ethics. As governments tighten rules in some regions and loosen them in others, companies must learn how to innovate without losing trust. The following section explores what compliance-by-design looks like and how leaders can turn regulation into a strategic advantage rather than a barrier.

NAVIGATE THE AI LAWS BEFORE THEY'RE WRITTEN

Artificial intelligence has triggered a global regulatory race as governments and international bodies craft frameworks to manage its transformative impact. For C-suite leaders of Global 2000 companies, mastering the regulatory landscape is now as critical as understanding the technology itself. The horizon bristles with new laws, guidelines, and ethical debates that will directly shape how organizations deploy AI systems. This complex and fragmented environment demands proactive navigation to ensure compliance while maintaining competitive advantage.

The European Union remains the regulatory pacesetter with its comprehensive EU Artificial Intelligence Act. This landmark law (one of the first broad frameworks regulating AI by risk category) entered into force in 2024 and begins phased implementation through 2026–2027. The Act classifies "high-risk" AI use cases such as recruitment, credit decisions, medical devices, and law enforcement, imposing strict requirements on providers. These include robust risk management protocols, high-quality datasets to minimize bias, and meaningful human oversight. An AI system used for hiring or college admissions in Europe must undergo rigorous testing and documentation to prove it avoids discrimination. Transparency is non-negotiable: systems interacting with humans must disclose their artificial nature, and outputs like deepfake videos require clear labeling. The Act's extraterritorial reach means any company offering AI-driven products or services in Europe must comply, regardless of headquarters location, effectively setting global benchmarks.

In 2025, the United States has moved in the opposite direction. The administration replaced previous safeguards focused on civil rights, privacy, and worker protection with the new "Removing Barriers to American Leadership in Artificial Intelligence" order. Federal procurement now emphasizes "Unbiased AI Principles," requiring vendors to certify that language models avoid ideological leanings, particularly DEI-related concepts. The National Institute of Standards and Technology has been instructed to revise its voluntary AI Risk Management Framework, stripping references to misinformation, DEI, and climate considerations. These mandates apply primarily to federal contracts, not the broader marketplace, signaling a shift from governance to growth.

Asia's regulatory landscape reflects diverse priorities. China continues to tighten oversight while pursuing its 2030 AI leadership goal. Recent rules require providers of recommendation algorithms and generative AI to register with authorities and comply with content guidelines aligned with "core socialist values." Strict data laws (including the Data Security Law and Personal Information Protection Law) limit how training data can be used. Global firms operating in China must navigate these constraints alongside geopolitical tensions and export controls. India has launched its national AI mission, building a compute platform targeting more than 10,000 GPUs and introducing draft guidelines for responsible AI. Japan and South Korea maintain soft-law approaches, issuing non-binding ethical principles focused on human-centricity and privacy. Singapore continues to lead with innovation-friendly frameworks, offering regulatory sandboxes for controlled experimentation.

Ethical challenges increasingly drive self-regulation as companies recognize that proactive standards protect both compliance and reputation. Algorithmic bias remains a top concern: hiring tools that inadvertently filter candidates by gender or lending models that disadvantage minority groups can trigger lawsuits and public backlash. Real-world cases underscore the risk. To mitigate, conduct regular bias audits, assemble diverse development teams, and deploy bias mitigation toolkits. Some jurisdictions mandate these practices, such as New York City's requirement for annual audits of AI hiring tools, which progressive firms adopt globally rather than limiting compliance to local laws.

Transparency and explainability have become essential for trust. Customers, employees, and regulators increasingly demand clarity on how AI systems make decisions. In high-stakes contexts like disease diagnosis or loan approvals, black-box models are unacceptable. Europe enshrines a "right to explanation" in certain cases, but expectations often exceed legal mandates. Companies invest in explainable AI techniques and maintain internal registries documenting each algorithm's purpose, data sources, and decision logic, practices that serve governance needs and reassure stakeholders.

Privacy and data governance remain foundational challenges as AI systems require vast datasets that often cross borders. Regulations such as the General Data Protection Regulation (GDPR) and the California Consumer Privacy Act (CCPA) impose strict rules on personal data use, and upcoming AI-specific requirements add obligations like documenting personal data usage in model training. Many enterprises adopt the strictest common denominator for data practices across jurisdictions, simplifying compliance while signaling commitment to privacy.

Fragmentation across jurisdictions creates implementation complexity. Establishing internal AI governance councils that set global standards while allowing local adjustments can help with this. These bodies typically include legal, compliance, IT, and business leaders who evaluate AI initiatives against unified ethical principles. Liability and insurance for AI systems are becoming more commonplace: when autonomous vehicles cause harm or AI-driven investment decisions fail, regulators must define accountability among manufacturers, operators, and possibly the AI systems themselves.

Looking ahead, regulatory development will accelerate. The EU AI Act is likely the first of many comprehensive laws. The UK continues to favor a principles-based approach, while U.S. Congress shows renewed interest in targeted legislation on facial recognition and social media algorithms. Asia will balance innovation with control, and Middle Eastern nations are drafting governance frameworks alongside massive AI investments.

By combining robust internal governance with active external engagement, organizations build resilience against regulatory risk while positioning themselves as trusted AI leaders in an era of heightened scrutiny. By combining robust internal governance with active external engagement, organizations build resilience against regulatory risks while establishing themselves as trusted AI practitioners in an increasingly scrutinized field. With regulatory guardrails taking shape, the next imperative is foresight: anticipating the technologies that will redefine competitive advantage before they hit the mainstream.

SPOT TOMORROW'S AI GAME-CHANGERS TODAY

Understanding emerging AI technologies is now a strategic imperative for those who want to capture opportunity and guard against disruption.

Today's business value comes from analytics and machine learning, but tomorrow's advantage will emerge from technologies still evolving in research labs and scaling through enterprise pilots. Leaders who anticipate these shifts can turn potential threats into durable competitive advantage.

Five categories of emerging AI are reshaping industries: generative AI and foundation models, multi-modal systems, autonomous intelligent agents, edge AI, and AI-augmented strategic decision support.

The following examples are drawn from publicly reported information and widely available sources:

Generative AI has emerged as one of the most immediately transformative technologies, as illustrated in public accounts such as Coca-Cola's "Create Real Magic" campaign. According to media coverage, this initiative—developed in collaboration with OpenAI and Bain—invited artists worldwide to remix brand assets using GPT-4 and DALL·E. Public reports highlight that the campaign generated thousands of AI-driven artworks showcased on Times Square billboards and significantly reduced creative production cycles from weeks to hours. Coca-Cola, as cited in industry news, now employs domain-specific models for marketing and demand forecasting, while maintaining a human-in-the-loop review process to protect brand tone and mitigate bias. Similarly, Pfizer's VOX generative AI platform, built on AWS Bedrock, has been publicly described as a tool that drafts scientific content and patent applications, saving thousands of hours annually and accelerating drug development timelines. These publicly documented cases demonstrate how generative AI has moved beyond novelty to operational impact, while also raising governance questions around intellectual property, ethics, and quality control.

Multi-modal AI is also redefining interaction, as reflected in public announcements about Google's Gemini models. These models integrate text, vision, audio, and code, enabling richer customer experiences and more complex reasoning. For example, public pilot programs in retail reportedly allow customers to upload product photos during support chats, enabling the AI to identify items, check warranty status, and provide spoken repair instructions. In healthcare, public case studies indicate that multimodal systems are combining imaging and clinical notes for diagnostic support, resulting in reduced interpretation errors.

Autonomous AI agents are progressing from concept to implementation, as seen in publicly available information about Siemens' introduction of Industrial AI Agents within its Xcelerator platform in 2025. Public reports indicate that these agents enable "automation of automation," orchestrating workflows, adjusting machine schedules, optimizing energy use, and reconfiguring production lines in real time. Early deployments at thyssenkrupp sites, as reported in the media, have shown productivity gains approaching 50 percent. Unlike static automation, these systems reportedly learn continuously and collaborate across digital and physical environments, signaling a shift toward Industry 5.0 where humans supervise adaptive AI ecosystems rather than micromanage processes.

Edge AI—which processes data directly on local devices like sensors or servers instead of sending it to a centralized cloud—has also gained traction, according to public sources. For instance, Walmart's global supply chain reportedly uses edge AI for inventory optimization and predictive maintenance in distribution centers. Public accounts highlight that by processing data locally, Walmart reduced stockouts by 30 percent and cut excess inventory by up to 25 percent, while federated learning ensures compliance with data residency laws. Similar public reports from the manufacturing and energy sectors highlight that edge deployments enable instant anomaly detection and autonomous control, balancing speed with privacy.

AI-augmented decision support is making its way into the boardroom, as evidenced by public disclosures from organizations such as JPMorgan Chase and Oracle. JPMorgan Chase reportedly integrates AI into risk management and financial planning through its LLM Suite and tools like Coach AI, which, according to public statements, helped advisors respond 95 percent faster during market volatility. Oracle's Fusion Cloud, as described in public materials, now embeds agentic AI for supply chain and ERP, automating scenario planning and exception handling. Public reporting emphasizes that these systems amplify judgment, surfacing insights from millions of data points and modeling geopolitical or market shocks in real time. Governance remains critical, as boards increasingly demand explainability and audit trails for AI-driven recommendations, as reflected in industry surveys and public statements.

Organizations that have been publicly recognized for success in this next wave of AI adoption share two traits: disciplined experimentation and workforce readiness. Public sources note that Walmart runs AI "sandboxes" to test models before scaling globally, while Pfizer partners with AWS to prototype generative AI use cases rapidly. Both companies, according to public reports, pair technology adoption with aggressive upskilling; Walmart's $1 billion investment in AI training has been widely covered. Industry surveys reported in the media indicate that over one-third of executives rank AI training as a top workforce priority, recognizing that human capability must evolve alongside machine intelligence.

Each technology carries risks like bias, opacity, security, underscoring the need for strong governance and ethical frameworks. Companies that innovate responsibly while scanning the horizon will convert disruption into advantage. With these technologies mapped, the next section examines how to prepare organizations for the structural shifts they will trigger across industries.

PREPARE FOR AI DISRUPTION BEFORE IT HITS

AI is driving business disruption faster than any previous technology because it enables new entrants to scale quickly while fundamentally changing how value is created.

Disruption now comes from multiple directions, and its pace is accelerating. Startups are not simply competing on cost or convenience; they are redefining what customers value. By embedding AI into their core offerings, these firms create experiences that feel intuitive, personalized, and immediate. This shift raises the bar for incumbents, forcing them to rethink long-standing assumptions about scale and differentiation. At the same time, customer expectations evolve rapidly as AI-powered services become the norm. What was once considered premium (e.g., instant recommendations, predictive support, frictionless transactions) quickly becomes baseline. Companies that fail to match these expectations risk losing relevance even if their traditional metrics look strong.

The ripple effects extend beyond direct competitors. Entire value chains are fragmenting as suppliers and partners deploy AI to expand their roles. A logistics provider that once focused on transportation can now offer predictive inventory management using machine learning. A component manufacturer can move upstream by using AI-driven design tools to co-create products with clients. These moves blur industry boundaries and erode the protective moats incumbents relied on for decades. In this environment, disruption is not a single event but a continuous process fueled by technology that accelerates learning and scales innovation at unprecedented speed.

Public reporting illustrates how quickly AI startups can challenge incumbents in banking. For example, in 2024, widely covered developments show Upstart and Zest AI captured significant U.S. market share by approving loans in minutes with advanced machine learning models, whereas legacy banks reportedly still relied on manual underwriting. According to public disclosures, JPMorgan responded by accelerating its own AI-driven credit risk platform, but the rapid shift highlighted in the media demonstrates how speed and personalization are redefining customer loyalty. Comparable patterns are reported in healthcare, where public case studies note that Tempus and PathAI use AI diagnostics to bypass traditional hospital networks, and in manufacturing, where media sources credit AI-native firms like Bright Machines with delivering customized production at scale.

To counter disruption, public sources indicate that Microsoft's venture arm and Google's Gradient Ventures monitor hundreds of AI startups globally, while BMW's Innovation Lab is reported to track autonomous systems and robotics to anticipate supply chain shifts. These scouting programs, according to industry reports, identify early signals—such as autonomous trucking—long before these trends reach mainstream headlines, enabling incumbents to test and learn ahead of competitors.

Media accounts note that Shell uses AI-driven simulations to model geopolitical and energy market shocks, preparing for futures where distributed AI optimizes local energy grids. Financial institutions, based on public reporting, run exercises imagining tech giants entering banking with AI-powered platforms, prompting defensive alliances and investments in differentiators that generalist entrants cannot easily replicate.

Organizational agility is frequently cited in public sources as the ultimate defense. For instance, Walmart's $1 billion investment in AI training and modular tech architecture, as widely reported, enabled the rapid rollout of generative AI for product descriptions and edge AI for inventory optimization when e-commerce rivals began using similar tools. According to public disclosures, Pfizer reserves a portion of its R&D budget for AI experiments—a strategy highlighted as critical when biotech startups accelerated drug discovery with generative models. Media reports emphasize that this foresight allowed Pfizer to integrate AI-driven approaches into ongoing projects, compressing timelines without compromising regulatory rigor. Key enablers of organizational agility, as described in industry surveys and public statements, include modular structures with empowered teams, dedicated innovation investments, data-driven decision processes, and continuous learning cultures.

Self-disruption is also increasingly common, according to public sources. Media coverage notes that Allianz launched a digital-only insurance brand using AI for instant claims and dynamic pricing—even though this move cannibalized parts of its traditional business. Public reports indicate that auto manufacturers like GM and Volkswagen acquired autonomous driving startups to secure future relevance, while retailers such as Walmart and Carrefour purchased AI-powered e-commerce platforms to strengthen omnichannel strategies.

Customer focus anchors all these moves, as emphasized in public materials. AI wins by solving problems better through convenience, personalization, speed, and cost. Amazon's AI-driven recommendation engine and 24/7 Alexa support, according to media reports, set expectations that ripple across industries. As a result, incumbents are now deploying conversational AI and predictive personalization before challengers can exploit these gaps.

External disruption also comes from regulation and societal shifts. Europe's AI Act imposes strict compliance obligations, creating openings for agile players who master governance early. Open banking rules enabled fintech challengers; similar dynamics will follow as AI regulations reshape data access and accountability. Meanwhile, AI-driven local manufacturing and 3D printing threaten global supply chains, forcing companies to rethink sourcing strategies.

Collaboration often provides the strongest response when entire sectors face transformation. Automotive manufacturers formed alliances for self-driving standards and shared mapping data to counter tech giants' scale advantage. Financial institutions created blockchain and AI consortia to pool resources for compliance and infrastructure. These partnerships require careful navigation of competition laws but deliver collective resilience.

The most successful companies treat disruption preparedness as a continuous capability, not a one-time exercise. They combine horizon scanning, scenario planning, agility, customer engagement, and strategic collaboration into an integrated defense. By staying vigilant and adaptive, organizations can turn potentially devastating disruptions into opportunities for reinvention and growth. Companies that want to shape their competitive future must look beyond their own walls and influence the rules of the game. The next frontier of advantage lies in defining how AI is governed across industries, a role that rewards those who lead rather than follow.

SHAPE INDUSTRY STANDARDS TO CONTROL YOUR AI FUTURE

Will your organization comply with standards or help create them? Standards matter because they create consistency that accelerates adoption. They help companies avoid costly rework when regulations emerge. Most importantly, organizations at the standards table can steer guidelines toward approaches that align with their technologies and strategies.

Influence flows through multiple channels. Formal standards bodies such as the International Organization for Standardization (ISO) develop technical specifications through working groups where companies send experts. Industry associations craft sector-specific guidelines that often become de facto standards. Government consultations shape emerging regulations. Open-source projects and published best practices establish informal norms that can be just as powerful.

ISO and the International Electrotechnical Commission (IEC) lead formal technical standardization through their joint AI committee. Companies like IBM, Microsoft, and Siemens actively participate in defining terminology, architectures, and testing methods. These standards influence global AI development and often migrate into regulatory frameworks. Participation offers early visibility and the chance to advocate for interoperability standards that enable a company's architecture. Early alignment reduces compliance costs and positions products for broader adoption.

Industry associations provide another avenue. Financial regulators such as FINRA issue AI guidance shaped by input from major banks. Automotive groups like SAE International define safety levels for autonomous vehicles. Healthcare alliances establish reference algorithms for diagnostic AI. Active participation ensures guidelines reflect operational realities. When the Healthcare Alliance for AI Diagnostics published standards for imaging algorithms, companies like GE Healthcare and Philips helped shape requirements that improved interoperability across hospital systems.

Government initiatives increasingly seek industry input. The European Commission consulted extensively with companies while drafting the AI Act. In the U.S., NIST developed its AI Risk Management Framework through industry workshops. Tech executives testify before legislative bodies, and global firms maintain policy teams to coordinate positions and engage regulators. This involvement helps create practical rules that protect society without stifling innovation.

Open standards and shared practices gain influence through voluntary adoption. Google's release of TensorFlow shaped machine learning globally. Similarly, Microsoft's Responsible AI Standard, published publicly, influenced ethics frameworks across sectors. An insurance company that publishes its AI accountability framework can see competitors adopt it, turning a proprietary approach into an informal industry standard. White papers and conference presentations amplify these effects.

Ethical leadership often drives self-regulation ahead of formal mandates. When Delta Airlines committed to transparent AI pricing policies, competitors faced pressure to match or risk reputational damage. Market forces can propagate best practices faster than regulation.

Certification programs create measurable standards for AI quality and compliance. Industry certifications, similar to ISO quality standards, reassure customers and partners while creating competitive advantages for certified organizations. If major players require certifications from suppliers, they become practical requirements across supply chains. This mechanism allows industries to self-regulate while maintaining flexibility that rigid regulations cannot offer.

When Siemens implemented rigorous AI validation to meet regulatory expectations, it also improved model performance. Publishing these outcomes encouraged competitors to adopt similar practices, effectively standardizing validation methods. Data-backed success stories carry more weight than theoretical guidelines.

Pre-competitive issues such as safety, ethics, and infrastructure benefit all players when addressed collectively. The automotive industry exemplifies this approach: manufacturers formed alliances to develop autonomous driving standards and share mapping data, recognizing that common foundations serve everyone. Financial institutions have created similar consortia for AI fraud detection.

Companies typically advocate for standards in foundational areas that do not compromise competitive advantages. They differentiate through implementation while supporting common baselines that expand markets. A firm might help establish explainability standards ensuring all companies provide basic transparency, then compete by offering superior explanation interfaces beyond minimum requirements.

Organizations need clear positions on key issues developed through cross-functional teams combining technical, legal, and business perspectives. Dedicated resources support participation in standards bodies and maintain consistent presence. Some companies fund university research centers or industry initiatives as part of their influence strategy. Employee participation requires time allocation and recognition as valuable work. Relationships with regulators and standards bodies need ongoing cultivation.

In 2023, the Open Compute Project (OCP) brought together companies such as Meta, Microsoft, and Intel to develop common specifications for AI hardware used in data centers. Their unified approach created interoperable designs for cooling systems and chip architectures, which regulators later referenced in energy-efficiency guidelines. This collaboration not only reduced costs across the ecosystem but also accelerated adoption of sustainable AI infrastructure. The outcome illustrates how coordinated standard-setting can align competitive interests with industry-wide progress.

Shaping industry standards secures the rules of the game, but true market dominance calls on leaders who can translate technological disruption into strategic advantage and cultural transformation.

LEAD WITH AI TO DOMINATE YOUR MARKET

Leadership in an AI-dominated world requires a fundamental shift from traditional executive capabilities to a new synthesis of technological fluency, ethical grounding, and human-centric vision.

Organizations must evolve their approaches to guide employees through continuous technological change while maintaining core values and purpose. Success depends on integrating AI strategy with business strategy versus treating it as a separate initiative.

Effective leadership in this new landscape rests on five essentials:

- Holistic vision that integrates AI into business objectives.
- Technical literacy: comprehension, not coding.
- Human-centered leadership that empowers and reskills.
- Balanced decision-making combining AI insights with human judgment.
- Ethical stewardship that turns responsibility into competitive advantage.

According to public reports and widely available sources, a holistic vision and strategy can transform AI from a cost center into a value creator. Leaders are advised to articulate how AI delivers value to customers and stakeholders while ensuring alignment with the company's mission. For example, public accounts describe how retail CEOs are envisioning AI-driven personalized experiences at scale, then mobilizing investments, partnerships, and talent to achieve those outcomes. Such vision incorporates both the opportunities and responsibilities associated with AI. Publicly, leaders have clarified what AI will and will not do; in one often-cited example, JPMorgan Chase has committed to automating routine customer interactions using AI while preserving human involvement for sensitive decisions like hardship loans. This level of clarity, as reported, reduces anxiety and fosters trust. Setting bold goals together with clear ethical boundaries, as reflected in public statements, helps organizations navigate transformation with confidence.

Publicly available information also highlights that technical literacy is now considered an essential leadership qualification, even for those without coding expertise. Similar to financial literacy for non-accountants, a basic understanding of AI enables executives to make informed decisions. Many companies, as documented in public reports, provide AI training for leadership teams or appoint technology experts to their boards. Leaders are now expected to look beyond budgets and timelines, asking about data quality, bias mitigation, and scalability. For instance, it has been publicly reported that Microsoft's board invested in structured AI education before approving major generative AI initiatives, ensuring that decisions were based on facts rather than hype. This growing emphasis on technology fluency, as seen in public disclosures, allows leaders to bridge business and technical perspectives, which is crucial for successful AI integration.

Human-centered leadership ensures AI augments rather than alienates people. Even with expanding AI capabilities, humans remain central as technology creators, supervisors, and beneficiaries. Leaders must communicate that AI enhances work rather than replacing human worth. Investment in reskilling programs demonstrates commitment to employee development. Celebrating human-AI collaboration success stories reinforces positive narratives. Innovation programs inviting employee AI project ideas tap internal creativity while building engagement. When AI displaces specific roles, ethical leadership provides transparency and support through retraining, role transitions, or generous severance. This approach builds loyalty and security that encourages broader AI adoption.

Continuous learning starts at the top. CEOs who openly discuss taking AI courses or visiting research labs model organizational learning culture. Their visible engagement with new knowledge encourages similar behavior throughout companies. Leaders become chief AI ambassadors whose attitudes toward learning cascade through organizations. Dismissive or fearful leadership breeds resistance, while curious and humble leadership fosters adaptation.

Decision-making evolves to balance AI insights with human judgment. AI generates vast analytics, predictions, and recommendations that can overwhelm executives. Leaders must develop data-informed decision approaches without surrendering to analysis paralysis. AI provides input among many factors requiring human wisdom.

According to public reports and widely available sources, Schneider Electric illustrates this approach through its global reskilling programs and internal AI innovation challenges, where employees are encouraged to pitch AI-driven ideas. These initiatives turn fear into creativity and engagement. Brand implications, employee morale, and strategic intuition often lie beyond AI's analytical scope. Leadership becomes dialogue between human experience and AI rationality. AI-using managers will replace those who don't. Supply chain executives leveraging AI forecasts outperform those relying solely on intuition, provided they understand AI guidance and can override it when necessary.

Ethical leadership establishes AI principles from the top. Leaders champion fairness, transparency, and accountability by embedding these values into AI deployment. This might involve ethics committees with board oversight or explicit ethical review steps in every AI project. Difficult decisions require moral courage. Leaders sometimes delay profitable AI deployments to address ethical concerns. Long-term reputational capital from responsible behavior typically outweighs short-term gains. Boards increasingly treat AI alignment with company values as governance responsibility. This extends to societal impacts including job displacement, privacy, and security. Leading companies engage proactively in these dialogues rather than reacting to criticism.

Performance leadership delivers results through AI-enabled transformation. Organizations leading in AI adoption show significant financial outperformance, achieving higher revenue growth and returns. This creates virtuous cycles where effective leadership drives AI success, validating strategic choices. Leaders set ambitious but realistic AI impact targets. Manufacturing executives might commit to doubling innovation rates through AI simulation. Media companies might target specific engagement uplifts from AI personalization. Progress requires measurement and communication. Companies increasingly include AI metrics in investor communications, demonstrating systematic integration into business success.

Resilient leadership prepares organizations for inevitable setbacks. AI journeys encounter failures, technological shifts, and external disruptions. Leaders must build organizations that absorb shocks and adapt quickly. Diversified AI initiative portfolios balance short-term gains with exploratory projects. When specific approaches become obsolete, alternative investments maintain momentum. This mirrors financial portfolio thinking applied to innovation. Crisis response demands decisive action, transparent communication, and rapid solution development.

Human elements of leadership grow more important as AI handles analytical tasks. Empathy, inspiration, and vision become key differentiators. AI can analyze engagement data, but only human leaders can have heartfelt conversations that truly motivate. Customer relationships and crisis management particularly benefit from human leadership. Genuine CEO communication during difficulties impacts trust far more than automated responses. Leaders serve as ultimate bridges between technology and humanity. As reported in public sources, Delta Airlines demonstrated this during the pandemic: CEO Ed Bastian's open and empathetic communication, according to widely available accounts, built a level of trust that a chatbot could not achieve.

Leading in AI-dominated markets demands unprecedented agility, knowledge, and ethical clarity. Yet AI also provides unprecedented leadership support and problem-solving capabilities when properly harnessed. The best leaders won't compute faster or memorize more data but will provide purpose, integrity, and unifying direction for human-machine organizations.

Success requires looking outward as much as inward. Organizations must engage globally, influence their environments, and lead within broader ecosystems. No company remains an island in the AI era. The most successful enterprises embrace global, ethical, and collaborative mindsets to harness AI for competitive advantage and societal benefit.

SUMMARY

Thriving in the age of AI requires executives to master both internal transformation and external navigation. Chapter 11 explored how organizations must look beyond internal transformation to navigate the global AI landscape, emphasizing the importance of understanding regional regulatory trends, anticipating technological disruption, and building organizational agility. The chapter highlighted the need for proactive compliance with evolving laws, participation in shaping industry standards, and the adoption of ethical AI practices to build trust and competitive advantage. It underscored the value of collaboration through alliances and consortia, and calls for a new style of leadership that blends technological fluency, ethical responsibility, and a human-centric vision to thrive in an AI-dominated world.

Chapter takeaways:

- Companies must actively navigate and shape the global AI landscape, leveraging regional strengths in technology, regulation, and talent to remain competitive.
- Proactive management of diverse and evolving AI regulations such as the EU AI Act, U.S. sectoral rules, and Asia's frameworks is essential for compliance and risk mitigation.
- Sustained advantage requires systematic scanning, experimentation, and adoption of emerging AI technologies, including generative AI, edge computing, and autonomous systems.
- Organizations must anticipate and counter AI-driven disruption from startups and new entrants by rethinking value creation, customer experience, and business models.
- Building modular structures, investing in workforce AI skills, and fostering a culture of rapid adaptation enable

companies to pivot quickly as AI capabilities and markets evolve.

- Active participation in developing technical, ethical, and industry standards allows firms to influence rules, ensure interoperability, and reduce future compliance costs.
- Implementing robust ethical frameworks covering bias mitigation, transparency, and human oversight builds trust, reduces reputational risk, and often shapes industry norms.
- Strategic alliances, consortia, and cross-industry partnerships accelerate AI learning, innovation, and resilience against sector-wide challenges.
- Effective AI-era leaders combine technical fluency, ethical judgment, and a human-centric vision to guide organizations through continuous transformation.
- Long-term success depends on orchestrating all these elements into a unified, adaptive approach: ecosystem engagement, regulatory strategy, technology foresight, disruption response, agility, standards, ethics, collaboration, and leadership.

As we transition to Chapter 12, we shift focus from establishing AI capabilities to sustaining them over time. The final chapter explores how organizations maintain momentum, adapt to continuous change, and evolve their AI strategies as technologies advance. You will discover frameworks for long-term AI success.

12

SUSTAINING MOMENTUM AND DRIVING FUTURE INNOVATION

The difference between sustained momentum and stagnation lies in how deeply you embed innovation into your culture, processes, and structure. Research tells a sobering story: 70 percent of AI project challenges stem from people and process issues, not technical ones. The most sophisticated algorithms become expensive paperwork without an organization culturally prepared to use them. Building continuous innovation requires intentional changes at three levels: culture, structure, and process.

According to publicly reported research from BCG, culture is increasingly recognized as the decisive factor in the success or stagnation of AI initiatives. Organizations that thrive with AI tend to foster environments where experimentation is encouraged and supported. For example, public sources highlight Google's widely known "20 percent time" initiative, which strategically invests in continuous innovation by allowing employees dedicated time for AI experiments.

Similarly, Walmart Labs has publicly shared its Tech Tuesday and AI Hackweek programs, designed to crowdsource automation ideas from associates and engineers throughout the company. These initiatives have led to rapid prototyping and piloting of projects like demand-forecasting tools and inventory-vision systems, embedding innovation into the company's operational DNA. The underlying message, as reported in various public forums, is that progress and creativity are prioritized over perfection. This approach to psychological safety forms the foundation for sustained innovation, enabling employees to take calculated risks and contribute to a dynamic, ever-evolving AI ecosystem.

Many enterprises establish AI Centers of Excellence (CoE) to institutionalize continuous development. These dedicated teams serve as innovation incubators, developing prototypes and spreading best practices throughout the company. A well-designed CoE prevents fragmented efforts and resource competition by centralizing knowledge and guidance. This structure creates the scaffolding for innovation. Break down silos by forming teams that blend data scientists, engineers, product managers, and business process owners. The hub-and-spoke model embeds AI specialists from the central CoE into business units and ensures capabilities spread organization-wide while maintaining consistency. Apart from allocating budgets, championing continuous AI innovation means articulating a compelling future vision and removing obstacles that stifle creativity.

Strategic planning cycles now include reviews of emerging AI opportunities as well as current projects. Product development roadmaps allocate capacity specifically for AI-driven enhancements. MLOps practices create pipelines for rapid experimentation, deployment, and iteration, preventing AI systems from growing stale.

The most effective organizations implement feedback loops that capture learning from each AI deployment. Post-project retrospectives ask: What worked? What needs improvement? These insights feed into a knowledge base that guides future innovation. Internal forums and communities of practice socialize successes and challenges across the organization, sparking new ideas through cross-pollination.

External environment monitoring has become a routine innovation practice. Leading companies designate teams to track AI research, startup activity, and industry trends. Some form university partnerships, join AI consortiums, or invest in startups to maintain cutting-edge awareness. Those alert to large language model emergence piloted generative AI use cases months ahead of competitors.

Staying ahead means maintaining readiness to move when new technology aligns with strategic needs. As one executive put it, borrowing from Eisenhower: "You can't predict which AI advancement will matter in three years, but by building the muscle for continuous innovation, you ensure your organization is ready to capitalize on whatever comes." Ultimately, the swarm is a commodity. Algorithms will become standardized. Compute power will be accessible to all. The only durable competitive advantage remaining is the human core. The organizations that win will be those that use the swarm not to replace their people, but to make their people smarter, faster, and more empathetic than their competitors. Once innovation becomes habitual, the next challenge is scale. Organizations must learn not only to generate bold ideas but to replicate and amplify them systematically across teams, functions, and markets.

SCALE FROM PILOT TO POWERHOUSE

Only 22 percent of organizations advance beyond proof-of-concept AI projects to generate real value. A mere four percent achieve substantial value at scale. Breaking free requires deliberate strategy that connects AI initiatives to business outcomes, builds robust infrastructure, and manages the human side of change.

The pattern is painfully familiar: initial enthusiasm, promising pilots, then a wall when it's time to operationalize broadly. Common culprits include disconnected strategy, insufficient infrastructure, talent shortages, and organizational resistance. Moving from isolated success to enterprise transformation demands a different approach.

Scaling succeeds when AI initiatives align tightly with strategic goals and core business processes. Pilots often emerge from innovation labs as tech-driven experiments, but to scale, business units must own and drive the AI agenda. Leaders who succeed concentrate resources on a few high-value domains rather than scattering efforts across dozens of use cases. According to public reporting, J.P. Morgan Chase established a cross-functional AI Center of Excellence that was specifically designed to support enterprise risk-reduction and operational-efficiency objectives. Reports indicate that by integrating data scientists, compliance officers, and product leads, this approach reduced fraud-detection cycle times by more than 40 percent and expanded AI-powered compliance checks across all major business units.

AI leaders tend toward aspirational targets that rally the organization and secure necessary investment. Every scaled AI project should tie to concrete KPIs such as reduced turnaround time, increased conversion rates, lower error rates. This linkage focuses implementation on what matters and earns stakeholder buy-in.

Pilots often succeed in isolation with ad-hoc data and tools. Scaling demands enterprise-wide infrastructure. Data silos kill scale. Organizations need integrated data platforms where customer, operational, and third-party data flow together. Modern architectures like cloud data warehouses provide the scalable foundation AI requires.

Successful scaling requires disciplined data governance. Different departments using inconsistent definitions or unreliable data will derail scale-out attempts. Establishing data governance bodies, enterprise data catalogs, and clear usage policies becomes critical infrastructure.

Instead of each team reinventing solutions, a centralized AI/ML platform enables rapid development with shared resources and best practices. Cloud infrastructure often provides unlimited compute scaling and global deployment capabilities. When one retailer created a centralized recommendation engine, every channel could access it without rebuilding the wheel.

Manual oversight during pilots won't suffice when AI runs thousands of decisions per minute. Organizations need frameworks that monitor performance, manage risks, and ensure compliance at scale. This starts with clear ownership: who "owns" each production AI model? Assign AI product owners within business units responsible for ongoing model success.

Evolve risk controls with scale. Regular model performance reviews catch drift before it causes damage. Automated monitoring tracks predictions against reality, triggering alerts when models veer off course. In regulated industries, governance ensures models meet compliance requirements before broad deployment. Ethics committees review high-impact systems for fairness and alignment with values.

Communication and change management determine adoption success. When an AI tool rolls out to hundreds of locations, success depends on training, documentation, and support channels. According to public reporting, Walmart's AI-driven inventory optimization program achieved success because store managers were trained to interpret system recommendations and make real-time order adjustments. Public sources also note that the company implemented structured feedback loops, allowing employees to flag anomalies in product demand and system outputs. This process helped refine the model and strengthen adoption across thousands of stores.

The hub-and-spoke model proves effective for cross-functional coordination. A central team develops core assets from initial pilots such as trained models, integration documentation, and change management playbooks. Then cross-functional teams customize and deploy solutions locally. This template approach enables learning from each deployment while maintaining consistency.

Portfolio management treats AI initiatives as investments to nurture, scale, or stop based on performance. Regular reviews ask: Which projects are ready to expand? Which need refinement? Which should pivot? Staged scaling (from one pilot to diverse test environments to full rollout) manages risk while maintaining momentum. After proving predictive maintenance in one factory, test in facilities across different countries with varied equipment. These phased deployments with clear checkpoints make scaling controlled rather than chaotic.

Scaling requires some standardization, which means local teams must adapt to enterprise solutions. Promoting "global over local optimization" helps teams understand that minor adjustments benefit the whole organization. Early involvement of end-users in pilot design creates co-ownership and reduces resistance.

Patience and persistence matter. Scaling takes longer than anticipated, integrating with legacy systems, training thousands of users, adapting to local conditions. Set realistic expectations with interim milestones. Celebrate progress to maintain momentum. Early adoption stages feel slow as foundations build, then adoption accelerates rapidly once critical mass is reached. Scaling technology is only sustainable when leadership scales with it. As systems grow more complex, leaders must evolve from project sponsors to architects of continuous transformation.

DEVELOP TOMORROW'S AI LEADERS TODAY

According to public reporting, automation and AI are projected to displace 92 million jobs by 2030, while creating 170 million new positions. Additionally, public sources indicate that nearly 40 percent of core job skills are expected to change within five years.

Long-term talent development has evolved from HR initiative to strategic imperative. The skills gap ranks as the number-one barrier to achieving AI ambitions. Success requires a three-pronged approach:

- Upskilling current employees

- Attracting specialized talent
- Creating a culture where continuous learning becomes second nature

Start with your existing workforce. Immediate priority centers on upskilling employees to contribute to and benefit from AI transformation. Technical staff need training in machine learning, data science, and MLOps. But equally critical is building AI literacy across the organization. Marketing professionals must learn to use AI segmentation tools. Operations managers need to interpret predictive analytics. Finance teams require skills to work with AI forecasting models.

Nearly half of employees want more formal AI training, viewing it as essential for success. Yet many feel unsupported by their employers. This gap represents both risk and opportunity. Structured programs ranging from workshops to certification courses should become part of performance development plans. Launch enterprise-wide AI academies where all levels participate in curricula covering basics, case studies, and hands-on projects.

Reskilling goes deeper than upskilling. When AI automates entire job functions, affected employees need pathways to new roles. Consider administrative staff whose tasks become AI-driven. With proper reskilling, they transition to analytical roles overseeing AI systems or handling exceptions the technology can't manage. Facilitate these transitions using multi-month training programs or rotational assignments. Over 80 percent of leaders struggle to find needed AI talent externally. The pipeline isn't large enough, and competition is fierce. Reskilling internal employees often proves the smartest path; they bring invaluable institutional knowledge that external hires lack.

As AI handles routine work, uniquely human capabilities become more valuable. Critical thinking, complex problem-solving, and creativity rise in importance. Employees spend more time on tasks requiring judgment, intuition, and interpersonal connection. Development programs should nurture these durable skills alongside technical training. Consider incorporating design thinking, ethical reasoning, and change agility into your AI training curricula.

Based on a December 2024 survey of 393 vice presidents and directors, three out of five VPs and directors have never attended AI training. Only 42 percent feel confident using AI tools without compromising data. Two-thirds feel ill-equipped to assess vendor offerings. These statistics reveal a critical vulnerability: leaders steering AI initiatives without sufficient understanding.

Executive education on AI fundamentals, capabilities, and risks has become essential. Short programs, workshops, and enrollment in "AI for Executives" courses help leaders ask the right questions and make informed decisions. When leadership becomes AI-conversant, they set realistic expectations, provide better oversight, and champion continuous learning throughout the organization.

Despite upskilling efforts, organizations need external AI specialists: data scientists, ML engineers, AI ethicists. Competition for these professionals remains intense. Attracting top talent requires positioning your company as an exciting place to work on meaningful AI projects. Large enterprises can leverage their data resources and scale to offer opportunities startups can't match. Retention matters as much as recruitment. AI experts value continuous learning; the field evolves rapidly. Providing access to research, conferences, and challenging problems helps retain talent. Creative approaches help too: university partnerships tap graduate pipelines, AI competition sponsorships identify promising individuals, and acqui-hiring brings entire teams aboard.

Most critical for long-term success: instilling a mindset of lifelong learning. With skill half-lives shrinking, employees must continuously adapt. Encourage this through learning stipends, dedicated development time, and recognition for skill building. Some set learning objectives and key results (OKRs): team goals for courses completed or certifications earned. The payoff is organizational adaptability. When employees habitually learn, they adopt new tools faster than competitors. Internal mobility (where employees move into roles requiring different skills) signals successful reskilling. As Adam Grant said, "In your career, the most valuable currency is not how much you know. It's how well you learn. In a stable world, success depended on building expertise. In a changing world, it hinges on evolving expertise. Potential is no longer defined by ability. It's a function of agility."

The next five to ten years will see new positions combining domain expertise with AI savvy. Marketing AI strategists who understand campaigns and data science. Manufacturing AI leads who bridge industrial engineering and AI-driven robotics. These hybrid roles become invaluable for implementation. Career paths must evolve accordingly. Today's entry-level analyst might progress to managing AI tools, then become an AI-enabled consultant guiding others, eventually reaching an AI product owner role. Organizations should map these new progressions, helping employees envision their future in an AI-augmented workplace. When companies invest visibly in employee skills, it builds trust and reduces fear. Establish mentorship programs pairing tech-savvy employees with others. Encourage grassroots learning communities where employees share AI experiences across departments.

With a skilled and adaptable workforce in place, organizations gain the human capital needed to navigate rapid change. But the organization itself must become agile enough to pivot when conditions shift. The next section explores how to build structures and mindsets that thrive on change rather than resist it.

ADAPT FASTER THAN THE MARKET

"Plans are worthless, but planning is everything" – President Eisenhower's wisdom rings truer in the AI era than ever before. The business environment has evolved from complex to truly chaotic. AI breakthroughs emerge with dizzying speed, regulations shift overnight, and competitors leverage AI in ways that reshape entire industries. Organizations that can't pivot quickly will find themselves obsolete, regardless of their current AI capabilities.

The VUCA world (Volatile, Uncertain, Complex, Ambiguous) is the new reality. Gen-AI models transformed entire sectors in months, forcing companies to react or risk irrelevance. Below are some best practices to embrace agility as a core competency.

Extend agile beyond IT. Most enterprises confine agile practices to software development. True organizational agility demands enterprise-wide adoption. Teams must rapidly form, execute, and dissolve in response to emerging opportunities. Decision-making should flow to those closest to the action, guided by clear principles rather than rigid hierarchies. Quarterly OKRs replace annual plans. Cross-functional squads tackle key initiatives without bureaucratic delays. Some companies implement scaled frameworks like SAFe, but the specific methodology matters less than the outcome: short feedback loops, iterative progress, and continuous adaptation.

Make small bets, learn fast. The pilot-and-scale approach becomes a broader agility mechanism. Instead of betting everything on one massive plan, organizations run controlled experiments, learn rapidly, and scale successes. When considering AI-driven customer service transformation, test in one region first. Gather data, refine the approach, then expand. This iterative method seems obvious, yet many organizations still default to all-or-nothing implementations.

Scenario planning beats prediction. The future of AI branches into multiple possible paths: steady evolution, radical transformation, or regulatory constriction. While we can't know which scenario will unfold, planning for each builds organizational flexibility. What happens if a breakthrough makes your industry's entry barriers disappear? How would you respond to regulations that suddenly constrain AI usage in core processes? By thinking through extremes and possibilities, organizations develop mental agility. You might never face the exact scenario planned, but the planning process creates readiness. Maintain a "backlog" of strategic responses ready for deployment when reality demands.

Decentralize decision authority. Adaptable organizations push decisions to the edges. Those closest to customers, operations, or emerging technologies often spot changes first and understand implications best. Set clear guardrails (i.e., budget limits, risk thresholds, brand guidelines) then grant teams autonomy within those boundaries. A front-line team noticing shifted customer behavior can adjust AI parameters immediately rather than waiting for executive approval. Mid-level managers can pilot new AI tools without lengthy proposals. This distributed authority accelerates response time dramatically. During COVID-19, companies that trusted local managers to adapt quickly outperformed those requiring corporate approval for every change.

Structure for flexibility. Some organizations restructure into networks of smaller units maintaining startup-like agility. While large enterprises can't become flat startups, they can move structurally toward greater flexibility. Reduce management layers to shorten feedback loops. Create multidisciplinary teams aligned to products or customer segments. Establish internal venture units operating under different rules to pursue breakthrough innovations. Organizational ambidexterity maintains efficiency in core operations while fostering exploration in other areas. Knowledge flows between stable and experimental units, preserving adaptability even as the company scales.

Embrace intelligent failure. Agility requires viewing failure as learning opportunity. When AI deployments underperform, dissect the experience: What assumptions proved wrong? What can we adjust? This requires psychological safety: employees must feel secure questioning results and acknowledging mistakes. Some companies simulate failures deliberately. If AI handles loan approvals end-to-end, maintain human capability as backup. Like pilots training for manual flight when autopilot fails, this redundancy provides resilience during transitions.

Build sensing mechanisms. Adaptation requires knowing when and what to adapt. Leverage data and AI itself to monitor their environment continuously. Use AI tools for competitive intelligence, sentiment analysis, and trend detection. Establish real-time dashboards tracking key metrics. When patterns deviate, teams can respond rapidly. External networks provide early warning systems. Partnerships with startups, customer advisory boards, and industry consortiums feed insights about emerging changes. The wider your sensing net, the earlier you detect shifts requiring adaptation.

Balance efficiency with resilience. Pure efficiency optimization creates brittleness. Accept some redundancy as the price of flexibility. Running two AI models that cross-check each other costs more than one, but provides insurance against failure. Maintaining human oversight might slow processes slightly but prevents catastrophic errors. Consider this trade-off as resilience investment. The cost of downtime or major incidents exceeds ongoing investment in adaptive capacity.

Cultural transformation enables structural agility. Employees comfortable with predictability must embrace constant change. Model this by treating strategy as living document rather than stone tablet. When plans need adjustment, explain clearly without defensiveness. Celebrate teams that pivot successfully, not just those who execute original plans perfectly. Communicate change as opportunity for growth rather than threat to stability. When employees see leaders learning and adapting, they're more likely to follow suit. Over time, change becomes routine rather than disruption.

Agility helps organizations pivot toward opportunities and away from threats. But what happens when unexpected shocks hit: system failures, cyberattacks, or regulatory upheavals? The next section explores building resilience that keeps organizations standing when AI elements fail or face unprecedented challenges.

DESIGN FOR DISRUPTION

An AI-resilient organization anticipates that the swarm will eventually falter. As companies embed AI deeper into critical operations, they create new vulnerabilities alongside new capabilities. Technical failures, adversarial attacks, and data poisoning are no longer just IT problems; they are existential threats to the swarm's integrity. True resilience requires accepting a fundamental truth of the AI era: efficiency is not the only metric. In fact, hyper-efficiency creates brittleness. To survive the inevitable shocks of a digital ecosystem, organizations must design for disruption by prioritizing human agency as the ultimate fail-safe.

Pure optimization creates fragility because a system designed solely for speed lacks the slack necessary to absorb shocks. In a humanized swarm, redundancy is not waste; it is insurance. Organizations must maintain the ability to operate manually when the digital swarm goes dark. Just as hospitals test emergency generators, AI-enabled firms must periodically run analog drills where they operate critical functions without algorithmic support. This ensures that human institutional memory doesn't atrophy, keeping the workforce ready to take the controls when the autopilot fails. Furthermore, automated systems need undeniable kill switches. If an AI controlling a supply chain or financial model begins behaving erratically, there must be a clear, authorized protocol for humans to sever the connection immediately. This power to intervene must be explicit, protected, and culturally celebrated as an act of stewardship rather than insubordination.

Resilience is not just about robust code; it is about a culture of vigilance where employees serve as the organization's immune system. Frontline staff often spot hallucinations or anomalies long before central monitoring systems do. A resilient culture encourages analysts to question dashboard outputs that defy common sense and provides clear channels for reporting strange AI behavior. When a worker flags a potential error, they are not slowing down progress; they are protecting the organization's integrity. This vigilance extends to security, where the human element involves red teaming not just for security gaps, but for ethical gaps. Diverse teams should stress-test models to see how they behave under extreme pressure or malicious input, ensuring the system degrades gracefully rather than collapsing catastrophically.

When disruption hits, opacity is the enemy of recovery because black box systems that cannot be explained resist troubleshooting. Resilience requires investment in explainability tools that allow humans to see into the model's logic during a crisis. If an AI denies a loan or shuts down a production line, the human operator must be able to interrogate the why. Building glass boxes means ensuring that documentation and data lineage are accessible. This transparency empowers humans to diagnose issues rapidly, turning a potential disaster into a manageable incident.

We are entering the era of 'Chaos Engineering.' This involves deliberately breaking your own AI systems to see what happens, essentially paying your engineers to act like gremlins. It sounds counterintuitive, but it's better to have your own team break the system on a Tuesday morning than to have a hacker do it on Black Friday. Think of it as a fire drill, but with actual fire. Controlled fire. Hopefully. Regardless of the technical sophistication, the core principle remains that the machine is designed for the probable, but the human is designed for the possible. By balancing algorithmic speed with human oversight, organizations build a resilience that can withstand the shocks of the future. The cost of this redundancy is dwarfed by the cost of a major failure, proving that the most resilient AI system is one that knows when to defer to its human partner.

Resilience protects organizations from shocks and failures. But sustainable AI success requires maintaining trust and legitimacy through ethical practices. The next section explores how continuous commitment to responsible AI becomes both shield and competitive advantage.

MAKE ETHICS YOUR COMPETITIVE EDGE

Ethical AI practice isn't a compliance checkbox, it's an ongoing commitment that defines whether your AI initiatives build or erode trust. Companies that treat ethics as a one-time exercise face predictable outcomes: legal penalties, reputation damage, talent exodus, and lost customer confidence. Those who embed ethics into every stage of the AI lifecycle convert responsibility into competitive advantage.

The ethical landscape shifts constantly. Face recognition technology went from widespread deployment to corporate retreat as privacy concerns mounted. Public sentiment and regulations evolve rapidly. What's acceptable today may face backlash tomorrow. Only continuous ethical vigilance ensures AI remains an asset rather than liability.

Leading companies establish AI ethics committees combining diverse perspectives such as data scientists, legal experts, business leaders, external advisors. These bodies review high-impact projects, guide ethical dilemmas, and set standards. Their role extends beyond approval to monitoring deployed systems and respond to incidents.

Some organizations appoint Responsible AI officers, analogous to Chief Privacy Officers but focused on AI ethics. They maintain AI inventories, ensure documentation exists, and lead training programs. They track global regulatory changes and adjust company guidelines accordingly. This dedicated oversight prevents ethical considerations from falling through organizational cracks.

Monitor bias continuously. Fairness isn't achieved once - it requires ongoing vigilance. Track AI outcomes across demographic groups. Measure error rates, approval rates, and other key metrics for disparate impact. When issues surface, have procedures ready: retrain models with better data, adjust algorithms, or temporarily revert to simpler approaches.

Diverse development teams catch blind spots homogeneous groups miss. Long-term hiring that brings varied backgrounds into AI design functions as preventive medicine. External stakeholder consultation (reviewing healthcare AI with patient advocates, for example) provides perspectives internal teams lack.

While complex models resist interpretation, organizations can still provide meaningful explanations. When AI declines a loan application, offer understandable reasons derived from the model. Techniques for generating explanations continue advancing, and regulatory pressure for transparency increases. The EU AI Act emphasizes explanation rights for high-risk AI decisions.

Some companies watermark AI-generated content or label AI-assisted interactions. Publishing annual AI use reports demonstrates accountability. Leading organizations share both successes and challenges, building credibility through honesty.

Responsible AI intertwines with data responsibility. Personal data used in AI requires careful handling: anonymization, proper consent, limited use scope. Privacy laws proliferate globally. Companies adopting the strictest standards as baseline simplify compliance across jurisdictions.

Advanced techniques like federated learning or differential privacy enable AI learning without exposing raw data. Early investment in privacy-preserving AI positions companies ahead of tightening regulations. Those treating privacy as afterthought face expensive retrofitting or system abandonment.

The EU AI Act represents the first comprehensive AI law, categorizing uses by risk level and imposing graduated requirements. High-risk applications face demands for risk assessments, documentation, transparency, and human oversight. Other jurisdictions develop their own frameworks. Global companies must adapt to this regulatory patchwork.

Continuous practice means tracking legislation through dedicated legal counsel or Responsible AI teams. Update policies proactively rather than reactively. Engage with regulators and standards bodies; influence development while understanding direction. Companies already conducting internal audits prepare naturally for external requirements.

Publishing AI ethics principles means little without implementation. Regular training ensures developers and managers understand practical applications. Walk through real scenarios highlighting ethical pitfalls and mitigation strategies. Make ethics training periodic, not one-time.

Create safe channels for ethical concerns. Engineers pressured to deploy questionable models need secure ways to escalate issues. This parallels broader whistleblower protections but focuses specifically on AI ethics. When employees see concerns addressed seriously, ethical culture strengthens.

Engage stakeholders continuously. Those affected by AI systems (i.e., customers, employees, communities) provide essential feedback. Post-deployment monitoring should include user experience alongside technical metrics. When patterns of frustration or unfair treatment emerge, they guide improvements.

Regular consultation with civil society representatives demonstrates commitment and provides external perspective. This engagement proves especially valuable when incidents occur. Established relationships and demonstrated good faith ease crisis resolution.

Ethical constraints often spur creative solutions. Companies developing targeted advertising without invasive data collection discover privacy-preserving techniques that become competitive advantages. Accessibility requirements lead to interfaces benefiting all users.

Visionary leaders frame ethics as innovation driver, not innovation inhibitor. They ask teams to solve problems within ethical boundaries, fostering creativity. Over time, this approach yields solutions that are both effective and responsible, a combination increasingly valued by customers, employees, and investors.

The next 5-10 years will surface new dilemmas. Deepfakes and synthetic media create verification challenges. The boundary between human and AI work blurs, raising questions about attribution and authenticity. As AI integrates into physical systems such as robots and autonomous vehicles, safety and liability stakes escalate.

Companies with mature ethical frameworks adapt more readily to emerging issues. They have processes for identifying new ethical dimensions, forums for discussion, and mechanisms for policy evolution. Rather than scrambling to address each new challenge, they incorporate them into existing structures.

As AI ethics gains prominence, leadership in this area becomes brand differentiator. Surveys show consumers increasingly factor AI responsibility into purchasing decisions. Talented employees prefer companies with strong ethical commitments. Investors incorporate AI governance into ESG assessments.

Being known for trustworthy AI creates tangible value. When controversies arise, companies with established ethical track records weather them better. Past good faith efforts buy credibility during crises.

Strong ethics, resilient systems, and adaptive organizations all require one critical ingredient: visionary leadership. The final section examines how executives must evolve to guide their companies through an era where AI reshapes every aspect of business and society.

LEAD THROUGH THE UNKNOWN

Leadership in the final phase of AI transformation requires a fundamental shift from managing technological adoption to orchestrating perpetual evolution. The executives who thrive will not be those who simply execute a static strategy, but those who build an organizational metabolism capable of absorbing shocks and seizing opportunities that do not yet exist. The task is no longer to predict the future with certainty, but to construct an organization that is robust enough to succeed in multiple possible futures.

This demands a move from episodic decision-making to a posture of infinite flexibility. Leaders must act as architects of optionality, resisting the urge to bet the entire enterprise on a single technological trajectory. Instead, they must cultivate a portfolio of strategic bets involving gradual evolution, sudden breakthroughs, and collaborative developments to ensure the organization retains the maneuverability to pivot as the landscape shifts. By tracking emerging trends not as inevitable destinations but as potential scenarios, leaders position their organizations to thrive regardless of which specific AI capability comes to dominate.

In this environment, the leader's primary role evolves into the guardian of human intent. As the swarm of AI systems grows more autonomous and capable, the human leader must aggressively assert the purpose behind the process. It is insufficient to merely implement efficiency; the leader must ensure that the swarm serves a distinct human purpose, anchoring technological power in moral wisdom. This is the ultimate check against the drift of automation: a leader who refuses to let the metric become the mission, ensuring that the organization remains a vehicle for human flourishing rather than just algorithmic optimization.

Furthermore, the capacity to unlearn becomes as critical as the capacity to learn. The strategies that secured success in the digital era may become liabilities in the AI era. Leaders must model the courage to abandon successful but obsolete models, demonstrating that intellectual humility is a strategic asset, not a weakness. This involves creating a culture where shifting views based on new evidence is celebrated as agility rather than criticized as inconsistency. By visibly shedding old habits, leaders grant their organizations permission to evolve.

Ultimately, this era offers a rare opportunity to shape history. Ten years from now, the distinction between organizations will not be their access to computing power, but the quality of their navigation through this transition. The leaders who leave a lasting legacy will be those who balanced the scale of the machine with the depth of the human spirit, cultivating institutions that did not just survive the AI revolution, but defined it by proving that intelligent machines are most powerful when they amplify the best of human potential.

Sustaining AI momentum requires weaving together all these elements: innovation culture, scaling excellence, talent development, organizational agility, system resilience, ethical commitment, and visionary leadership. No single component suffices; together they create organizations capable of continuous evolution in an AI-dominated future. The journey doesn't end with implementing AI, it begins there. The companies that thrive will be those that treat transformation as an ongoing capability and not a destination to reach.

SUMMARY

Sustaining momentum in the AI era is about building the capability for perpetual transformation. Throughout this chapter, we've explored the essential elements that keep organizations moving forward: embedding innovation in your DNA, scaling from pilots to enterprise impact, developing tomorrow's talent today, maintaining agility amid chaos, designing for disruption, making ethics a competitive advantage, and leading through uncertainty with vision and courage.

These are interwoven threads that together create organizational resilience and adaptability. Innovation without ethics leads to backlash. Scaling without talent development creates brittle systems. Agility without resilience leaves you vulnerable to shocks. Leadership without all these elements lacks the foundation to succeed.

The companies that thrive in the next decade won't be those with the most sophisticated algorithms or the largest AI budgets. They'll be the ones that master continuous evolution, organizations where learning is constant, adaptation is reflexive, and human purpose guides technological power. As AI reshapes the landscape of work, the future of organizational change will be measured by how effectively we enable people to think with it. Change management will activate dynamic learning ecosystems, accelerate the movement of insight, and scale human interpretation faster than machines can mimic it. The organizations that thrive will be those that treat knowledge as a living system. Looking ahead, roles will evolve into sensemaking, abstraction, and the ability to recombine what AI surfaces into novel value.

Your Call to Action

As we close, we invite you to move from reading to doing. The frameworks and strategies in these pages mean nothing without implementation. Your organization's AI journey is unique, but the principles we've outlined provide a compass for navigation.

Start with these immediate steps:

- Assess your current state. Where does your organization stand on each dimension we've discussed? Are you strong in innovation but weak in scaling? Excellent at ethics but struggling with talent development? Identify your gaps and strengths honestly.
- Choose your first move. Don't try to transform everything at once. Select one area where improvement would have the

highest impact. Perhaps it's establishing an AI Center of Excellence, launching an enterprise-wide AI literacy program, or creating an ethics review board.
- Set a 90-day goal. Make it specific and achievable. "By [date], we will have [concrete outcome]." This creates momentum and demonstrates that change is possible.
- Engage your team. Share insights from this book with colleagues. Start conversations about what AI transformation means for your organization. Remember, sustainable change requires collective effort.
- Connect with the community. The challenges you face aren't unique. Other leaders are navigating similar territory. Join the conversation.

Stay Connected

The dialogue doesn't end with this book's final page. AI evolution accelerates daily, and we all benefit from shared learning. We invite you to:

- Connect with us on LinkedIn to continue the conversation
- Share your experiences as you implement these strategies
- Ask questions when you encounter challenges
- Celebrate successes as your organization achieves AI milestones

Find us at https://www.linkedin.com/in/arendschuring/ and https://www.linkedin.com/in/andyforbes/ where we regularly share insights on AI transformation and engage with leaders like you who are shaping the future of business.

A PERSONAL MESSAGE

Thank you for investing your time and attention in this journey. The fact that you've read this entire book signals your commitment to leading your organization thoughtfully through the AI revolution. That commitment matters - not just for your company's success, but for ensuring AI develops in ways that benefit humanity.

Every organization that approaches AI with wisdom, courage, and ethical grounding helps shape a better future. Your leadership ripples outward, influencing employees, customers, competitors, and communities. The choices you make about AI implementation set precedents and create possibilities.

THE ROAD AHEAD

We stand at an inflection point in human history. The next decade will see AI capabilities expand in ways we can barely imagine today. Some predictions will prove too conservative; others too ambitious. Certainty about specifics is impossible, but one thing remains clear: organizations led by thoughtful, adaptive, human-centered leaders help define the transformation.

The essence of this journey captures a critical truth: as AI systems grow more powerful and interconnected, resembling swarms of coordinated intelligence, our role isn't to resist or merely adapt using the playbooks of the past. It is to ensure these systems serve human flourishing—to actively humanize the swarm within an AI-first reality.

This requires the kind of leadership we've described: visionary yet flexible, bold yet ethical, technological yet deeply human. It demands organizations that can sustain momentum through continuous evolution rather than periodic revolution, accepting that the only stable state in this new era is one of constant, purposeful adaptation.

YOUR LEGACY

Ten years from now, people will look back at this moment as the beginning of the AI age. They'll study how different organizations responded, which ones thrived, which ones struggled, and why. Your actions today write that history.

Will your organization be remembered as one that seized the AI opportunity with wisdom and courage? As one that balanced innovation with integrity? As one that kept humans at the center while embracing technological transformation?

The answer lies in what you do next. The frameworks are in your hands. The future awaits your leadership.

APPENDIX: REFERENCES

Chapter 1: Embracing the AI Swarm

Chambers, J. (2015). *Keynote Address at BoxWorks & Cisco Live.* (Source of the "40% of companies will not survive" prediction).

Goldman Sachs Research. (2023). *Generative AI could raise global GDP by 7%.* (Data on productivity and decision cycle acceleration).

UPS Corporate Strategy. *ORION (On-Road Integrated Optimization and Navigation).* (Case study on task decomposition saving 10 million gallons of fuel).

Netflix Technology Blog. *Algorithms and Creative Decision Making.* (Shift from "rules-based" to "outcome-based" governance).

Microsoft WorkLab. (2024). *AI in Customer Service Case Studies.* (Farm Credit Canada & Estafeta reporting 78% reduction in routine tasks).

Ping An Insurance Group. *Annual Report.* (Documentation of "One-Click Claim" reducing processing to 10 seconds).

Amazon Web Services (AWS). *Moderna Case Study.* (70% acceleration in research time for COVID-19 vaccine).

Chapter 2: The Future of Work – A Task-Based Economy

LinkedIn Economic Graph. (2024). *Skills-First Report.* (Data confirming 25% skill shift since 2015).

McKinsey Global Institute. (2024). *The economic potential of generative AI.* (60–70% of work hours are automatable).

Goldman Sachs. (2023). *Global Economics Analyst.* (Two-thirds of U.S. occupations exposed to automation).

Anthem Health / X12 Standards. *EDI 837 Standard (2300 Loop).* (Technical basis for the "2,300 task" decomposition claim).

JPMorgan Chase. *Annual Report & Gloat Case Study.* (Talent marketplace facilitating 30,000 assignments/conversations).

MIT Project Nanda. (2025). *Enterprise AI Impact Report.* (95% failure rate of pilots due to organizational issues).

Chapter 3: Leading Organizational Transformation

Wang, D., et al. (2016). "Deep Learning for Identifying Metastatic Breast Cancer." *Harvard Medical School.* (The "Centaur" study: AI 92%, Human 96%, Combined 99.5%).

Google Research. *Internal Engineering Velocity Studies.* (Coding assistants increasing velocity by 35% and reducing bugs).

Chapter 4: Empowering Teams and Individuals

Dell'Acqua, F., et al. (2023). "Navigating the Jagged Technological Frontier." *Harvard Business School Working Paper.* (Consultants using AI were 25.1% faster with 40% higher quality).

Brynjolfsson, E., et al. (2023). "Generative AI at Work." *NBER Working Paper No. 31161.* (14% productivity boost in customer support).

Amazon. (2019/2024). *Upskilling 2025 Pledge.* ($1.2 billion investment in workforce retraining).

Workplace Intelligence & Writer. (2024). *State of AI in the Workplace.* (31% of employees admit to "sabotaging" AI tools).

Chapter 5: Human-Centered Change & Ethics

McKinsey Global Institute. (2024). *Global Skills Survey.* (87% of companies reporting skill gaps).

Obermeyer, Z., et al. (2019). "Dissecting racial bias in an algorithm." *Science.* (80% bias reduction by correcting proxy variables).

Angwin, J., et al. (2016). "Machine Bias." *ProPublica.* (COMPAS algorithm showing 45% false positive rate for Black defendants).

Chapter 6: Future Industry Scenarios

Future Today Institute. *Tech Trends Report.* (General reference for industry scenario planning).

Gartner. *Hype Cycle for Artificial Intelligence.* (Context for "trough of disillusionment" discussions).

Chapter 7: Frameworks for Execution

Boston Consulting Group (BCG). (2024). *AI Value Report.* (Only 4% of companies achieve substantial value).

McKinsey & Company. (2025). *State of AI.* (Only 1% of companies are "mature").

Chapter 8: Managing Resistance

Prosci. (2024). *Best Practices in Change Management.* (63% of organizations cite human factors as the top barrier).

Chapter 9: Scaling the Swarm

Bank of America. (2024). *Digital Banking Press Release.* ("Erica" surpassing 2.5 billion interactions).

Chapter 12: The Future of the Profession

Kearney. (2025). *CEO AI Readiness Study.* (Gap between 89% importance vs. 25% preparedness).

Mercer. (2024). *Global Talent Trends.* (58% of executives say tech outpaces retraining).

APPENDIX: THE BOOK, CONDENSED

THE WHOLE BOOK IN ONE SENTENCE

Traditional change management breaks down in an AI-first world, and organizations must redesign how change is sensed, shaped, and sustained when intelligence is embedded directly into work itself.

THE WHOLE BOOK IN ONE PARAGRAPH

This book argues that Organizational Change Management, as practiced for the last decade, was built for slow-moving, human-centered systems and is fundamentally mismatched to AI-driven environments where systems learn, adapt, and act continuously. In an AI-first organization, change is no longer a discrete event to be communicated and trained into compliance; it is a persistent condition that must be architected into roles, workflows, governance, and decision-making. The book reframes OCM from a downstream adoption function into an upstream design discipline, one that aligns incentives, accountability, human judgment, and machine capability from the start. Readers are guided toward a future where change is not managed episodically, but operationalized as a core organizational capability.

TEN SMART THINGS TO SAY WHEN TALKING ABOUT AI AND OCM

1. "In AI-first systems, change is continuous, so OCM has to be architectural, not episodic."
2. "You cannot train people into trusting systems that were poorly designed."
3. "AI does not eliminate change resistance; it exposes weak organizational design."
4. "OCM has to move upstream, into how work is structured, not just how it is explained."
5. "Adoption is a lagging indicator; alignment is the leading one."
6. "If the system learns faster than the organization, governance becomes the bottleneck."
7. "Human-in-the-loop only works when humans still own meaningful judgment."
8. "AI forces us to redesign roles, not just reskill people."
9. "Communication cannot fix misaligned incentives in AI-enabled workflows."
10. "The real OCM challenge with AI is not fear of machines, it is loss of clarity about responsibility."

ABOUT THE AUTHORS

Arend Schuring is a Prosci-certified organizational change management leader with more than two decades of experience guiding organizations through transformation. He holds a BBA from Baylor University in International Business and Performance Improvement Technologies, and is passionate about finding creative ways to engage people from all backgrounds and cultures, helping teams discover solutions and unleash their full potential.

Arend's projects have spanned a wide range of industries and technologies, reflecting his adaptability and breadth of expertise. Known for translating complex initiatives into clear, human-centered communications, Arend brings a rare blend of strategy and practical change execution to every engagement. His evidence-based, people-first philosophy has helped organizations turn change into a catalyst for growth and performance.

Andy Forbes is an information technology architect, strategist, and long-time technology leader with a career devoted to helping organizations deliver complex systems with clarity, discipline, and measurable impact. With more than four decades of hands-on experience in enterprise architecture, program management, and digital transformation, he has guided teams through challenging design and delivery decisions.

Andy has co-authored multiple books on AI and Salesforce delivery, including a guide on how modern project teams can use AI to accelerate design, configuration, development, and testing. His work extends beyond CRM: he has also written a growing collection of books on the practical application of AI across diverse real-world topics, from IT execution to physical fitness and home cooking, making advanced capabilities approachable for both technical and nontechnical audiences.

Throughout his career, Andy has focused on bridging strategy and execution, simplifying complexity, and empowering teams to leverage AI and modern delivery practices to work smarter and faster. He is passionate about helping organizations and people understand where technology is heading and how to prepare their people, processes, and architectures for an AI-driven future.